Remote Sensing Time Series Image Processing

T0187998

Taylor & Francis Series in Imaging Science

Series Editor

Qihao Weng

Indiana State University

Published Titles

Remote Sensing Time Series Image Processing
Qihao Weng

For more information about this series, please visit: www.crcpress.com

Remote Sensing Time Series Image Processing

Edited by
Qihao Weng

CRC Press
Taylor & Francis Group
Boca Raton London New York

CRC Press is an imprint of the
Taylor & Francis Group, an **informa** business

Contents

Preface vii
Acknowledgments xv
Editor xvii
Contributors xix

Part I   Time Series Image/Data Generation

1. **Cloud and Cloud Shadow Detection for Landsat Images:**
   The Fundamental Basis for Analyzing Landsat Time Series 3
   Zhe Zhu, Shi Qiu, Binbin He, and Chengbin Deng

2. **An Automatic System for Reconstructing High-Quality**
   Seasonal Landsat Time Series 25
   Xiaolin Zhu, Eileen H. Helmer, Jin Chen, and Desheng Liu

3. **Spatiotemporal Data Fusion to Generate Synthetic High Spatial**
   and Temporal Resolution Satellite Images 43
   Jin Chen, Yuhan Rao, and Xiaolin Zhu

Part II   Feature Development and Information Extraction

4. **Phenological Inference from Times Series Remote Sensing Data** 69
   Iryna Dronova and Lu Liang

5. **Time Series Analysis of Moderate Resolution Land Surface**
   Temperatures 89
   Benjamin Bechtel and Panagiotis Sismanidis

6. **Impervious Surface Estimation by Integrated Use of Landsat**
   and MODIS Time Series in Wuhan, China 121
   Zhang Lei and Qihao Weng

Part III   Time Series Image Applications

7. **Mapping Land Cover Trajectories Using Monthly MODIS Time**
   Series from 2001 to 2010 137
   Shanshan Cai and Desheng Liu

8. Creating a Robust Reference Dataset for Large Area Time
   Series Disturbance Classification 157
   *Mariela Soto-Berelov, Andrew Haywood, Simon Jones, Samuel Hislop,
   and Trung H. Nguyen*

9. A General Workflow for Mapping Forest Disturbance History
   Using Pixel Based Time Series Analysis 173
   Feng Zhao and Chengquan Huang

10. Monitoring Annual Vegetated Land Loss to Urbanization with
    Landsat Archive: A Case Study in Shanghai, China 205
    Qingling Zhang and Bhartendu Pandey

Index 221

Preface

A New Direction in Remote Sensing and Imaging Science

Great advances have occurred in numerous sub-fields of remote sensing and imaging science since the turn of the 21st century. Commercial satellites acquire imagery at a spatial resolution previously only possible with aerial platforms. Hyperspectral imaging collects detailed information of materials and estimates their abundance in the Earth's surface, enabling the use of remote sensing to replace data collection that was formerly limited to laboratory testing or expensive field surveys. Lidar technology can provide high-accuracy geometric information for man-made structures and natural landscapes. In addition, radar technology has been reinvented since the 1990s, due particularly to the increase of space-borne radar programs while more diverse applications are conducted using the technique of radar interferometry (Weng, 2012). In fact, these technologies have been integrated with more established aerial photography and multispectral remote sensing techniques, as well as with unmanned aircraft systems. These integrated systems have been a main stream of current remote sensing research and applications.

With the advent of the new sensor technology, the reinvention of "old" technology, and more flexible platforms and more capable computational techniques, the field of remote sensing is rapidly gaining, or regaining, interest in the geospatial technology community, governments, industries, and the general public (Weng, 2012). Today, remote sensing has become an essential tool for understanding the Earth and managing human-Earth interactions. Global environmental problems have become unprecedentedly important in the 21st century. Thus, there is a rapidly growing need for remote sensing and Earth observation technology that will enable monitoring of the world's natural resources and environments, managing exposure to natural and man-made risks and more frequently occurring disasters, and helping the sustainability and productivity of natural and human ecosystems (Weng, 2012). In this context, optimal use of temporal resolution/revisit has become an emerging trend, while the technology of satellite sensors continues to witness improvements in spatial and spectral resolutions. The improvement in temporal revisit relies mainly on the technologies of sensor, platform, and satellite systems, but enhancement of the temporal resolution of observation can also be achieved by using a long-term archived imagery (Zhang and Weng, 2016) and/or through image/data fusion of different sensors (Weng et al., 2014).

The improvement in temporal resolution/revisit would allow for the large accumulation of images for a specific location, creating a possibility for time series image analysis and eventual real-time assessment of scene dynamics.

The free access to long-term image depositories at coarse spatial resolution, such as MODIS, offers a great opportunity to understand how the Earth's surface is changing, to identify the causes and effects of the changes, and to predict future changes. However, time series image analysis did not become a new research frontier until the availability of free Landsat data from the United States Geological Survey (USGS) after 2008 (Woodcock et al., 2008). Studies that utilize medium-resolution time series imagery and produce consistent maps are highly valuable for monitoring land cover and land use changes (Sexton et al., 2013; Zhu and Woodcock, 2014), and to determine the impacts of urbanization on energy, water, carbon cycles, vegetation phenology, and surface climate (Weng and Fu, 2014). After considering internationally available medium-resolution satellite missions such as Sentinel-2, CBERS, and IRS, it is apparent that rich sources of Landsat-like data would allow us to monitor the Earth's surface from continental and global scales to regional and local scales (Gutman et al., 2008; Gao et al., 2012). Recently, USGS Landsat Analysis Ready Data (ARD) for the conterminous United States has become available for download (https://earthexplorer.usgs. gov). Landsat ARD makes the Landsat archive more accessible, easier to use, and will foster time series image analysis for various applications. It is expected that future researches will shift from the heavy labor investment in data procurement and preprocessing to better utilization of time series images for extracting, interpreting, analyzing, and modeling purposes. Furthermore, numerous computer codes, algorithms, systems, products, and publications are being created/published to support time series image processing and applications. To facilitate on-line and off-line learning and knowledge sharing, Indiana State University created a website in 2016 for sharing computer codes, algorithms, systems, products, and publications that support remote sensing observations and applications, digital image processing, and the extraction of geophysical and biophysical information. It is hoped that through co-learning, sharing, and collaborating, an e-community can be built among researchers, practitioners, teachers, and students, which is named: Remote Sensing E-community for Digital imaGe procESsing (RS-EDGES, http://rs-edges.net/dokuwiki/doku. php). By sharing codes, algorithms, products, and publications, researchers can disseminate their results more efficiently and learners can save much precious time by avoiding reinventing the existing techniques and models.

Synopsis of the Book

Methods and techniques of time series image analysis have been widely applied in topics ranging from vegetation dynamics to wetland, agricultural and range land, climate, hydrology, and urbanization. Wide spread interests on this topic have been demonstrated by a large number of published articles in peer refereed journals and by an increased number of presentations in professional conferences. The aim of this book is to bring together selected active, recognized authors in the field of time series image analysis and to present to the readers the current state of knowledge on remote sensing time series image processing and future directions. This book is intended for use by researchers and students in both remote sensing and imaging science. It may serve undergraduate and graduate students as a textbook for those majoring in remote sensing, imaging science, civic and electrical engineering, geography, geosciences, planning, environmental science, resources science, land use, energy, and GIS, but it can also be used as a reference book for practitioners and professionals in the government, commercial, and industrial sectors.

This book consists of three parts. Part I addresses methods and techniques for generating time series image datasets; Part II examines feature development and information extraction methods for time series imagery; and Part III illustrates various applications using time series image analysis. Three chapters are included in Part I. Chapter 1, by Zhu and his colleagues, investigates techniques for cloud and cloud shadow detection from satellite time series imagery. Cloud and cloud shadow detection is an essential pre-processing step for analyzing Landsat time series. This chapter provides a comprehensive review of existing algorithms, including physical-rules based and machine-learning based algorithms for single-date imagery, and those for multi-temporal images, better suited to remove thin clouds. This review provides guidance on the selection of cloud and cloud shadow detection algorithms for various applications, especially for using Landsat time series. In Chapter 2, the issues of clouds and cloud shadows, as well as the scan line corrector (SLC) problem of Landsat 7 are further addressed by developing an automatic system in order to generate high-quality time series imagery. This system is designed to interpolate contaminated pixels in all images of a time series based on the Neighborhood Similar Pixel Interpolator and an iterative process. The input data of the system are a time series of original Landsat images with cloud and cloud shadow masks, while the output is a time series without missing pixels caused by clouds, cloud shadows, and SLC-off gaps. To take advantage of different spatial and temporal characteristics of multi-sensor data, Chapter 3 introduces two spatiotemporal data fusion techniques, namely, NDVI Linear Mixing Growth Model (NDVI-LMGM) and Flexible Spatiotemporal DAta Fusion method (FSDAF). Both algorithms aim at generating synthetic time series data by fusing rich spatial details and frequent temporal information from different satellite sensors. While NDVI-LMGM is designed to construct high spatial- and temporal-resolution NDVI time series by using a spatial moving window to address local variability, FSDAF focuses on producing high spatial-resolution reflectance data by employing a thin-plate-spline interpolation method to deal with abrupt land cover type change. These spatiotemporal data fusion methods show a potential to improve the capability of monitoring rapid surface changes over heterogeneous landscapes.

Part II contains three chapters concerning feature development and information extraction from time series imagery. In recent decades, the applications of time series remote sensing data have increasingly addressed phenological phenomena, representing biological and environmental cycles in ecosystems functioning at landscape scales as well as signals of important climatic and ecological transitions. Chapter 4, by Dronova and Liang, presents a review of some key remote sensing-based phenological metrics, research approaches, and their major applications in ecosystem function and climate change studies. It starts by introducing the commonly used metrics of the single-season phenological trajectory of spectral indicators of vegetation greenness and schedules of their dynamics. It then discusses common applications of single-year phenology in studies of agricultural landscapes, forests, and complex hydrological ecosystems. Next, multi-year phenological analyses are reviewed including local-scale applications of *in situ* photographs (phenocam data), regional-scale analyses of trends in vegetation greenness and land cover dynamics, and integrative assessments of multi-year phenological trajectories to detect anomalies and/or directional ecosystem shifts. The final section concludes with a review of major applications at local, regional, and continental to global scales, key caveats to interpretations of remotely sensed phenology, and new computational possibilities to facilitate such studies in the future. Chapter 5 investigates land surface temperature (LST), a highly dynamic parameter that drives many terrestrial physical processes. However, the analysis of satellite LST is complicated due to cloud gaps and the pronounced spatiotemporal variability of LST. A prominent way to overcome these limitations is to model the annual LST cycle (ATC) through time series analysis by deriving annual cycle parameters (ACP). This chapter presents the ACP derived from MODIS collection-6 LST data for Europe and North Africa and introduces a new parameter, the coefficient of determination. Among the large number of potential applications, examples of ACP-based SUHI analysis and LST downscaling are presented. In Chapter 6, an innovative technique for extracting and mapping of impervious surfaces from satellite imagery is presented. Impervious surface data is important for analysis of urban environments, including biogeochemical cycles, climate change, biodiversity, and sustainable development. This chapter develops an efficient method for differentiating impervious surfaces from pervious surfaces based on temporal features derivable from Landsat and MODIS time series. The method uses dynamic time warping (DTW) distance to measure similarity of temporal features from reconstructed time series Biophysical Composition Index, and applies semi-supervised Support Vector Machine to map impervious surfaces at an annual frequency. The case study in Wuhan,

China, illustrates the spatio-temporal pattern of urbanization process from 2000 to 2015.

The four chapters in Part III demonstrate applications of time series image processing in land cover change, disturbance attribution, vegetation dynamics, and urbanization. In Chapter 7, Cai and Liu develop an integrated change detection and classification approach for mapping land cover trajectories using monthly MODIS time series. This approach first identifies stable periods in a time series based on a change date detection algorithm; a modified support vector machine method is then applied to adaptive time series for continuous classification. The method was experimentally used in a study area in southeast Ohio, USA, by constructing land cover trajectories at a 32-day interval using the MODIS time series between 2001 and 2010. The accuracy of trajectories yielded generally greater than 0.84 with an average accuracy of 0.863 in the period of observation. In Chapter 8, Soto-Berelovab and colleagues present a method for time series disturbance classification for a large-area reference dataset. In recent years, algorithms have been developed to detect spectral changes in pixels over time. However, in order to attribute spectral changes to disturbance agents (e.g., logging, fire), reference data are required. This chapter presents a practical framework for the creation of a reference dataset that takes advantage of an existing forest inventory plot network. The strength of this method lies in utilizing a sampling framework that is stratified across an entire jurisdiction using a fixed sampling design, and it is statistically robust. The created reference dataset consists of almost 8000 reference pixels over a large area in Victoria, Australia. A number of ancillary datasets and information are used by trained interpreters to attribute disturbance information. This reference dataset is then used in a machine learning environment to produce classified disturbance maps over a 28-year period. In Chapter 9, a method for producing forest disturbance history using pixel based time series analysis is presented. Under the North America Forest Dynamics (NAFD) study an automatic, efficient, and versatile workflow for evaluating forest disturbance history with Landsat observations has been developed. This approach has been applied to the conterminous US (CONUS) for the time period of 1986–2010. Comprehensive annual forest disturbance maps have been produced for 434 scenes covering the CONUS. Chapter 9 first reviews more than 350,000 Landsat scenes resident at the USGS EROS Landsat archive, and selected >24,000 good quality scenes for the study. The annual maps were merged into two 30-m spatial resolution summary US national disturbance maps, which covered the first and last disturbances (when more than one disturbance occurred) for nearly all US locations. These maps were produced using an updated version of the Vegetation Change Tracker (VCT) algorithm, an automated time series forest change analysis approach that utilizes temporally dense Landsat Time Series Stacks (LTSS) to map forest disturbance events. These maps are referred to as the NAFD-NEX dataset. The NAFD-NEX forest disturbance maps have been validated using a visual assessment of Landsat time series images and high resolution image observations. These maps are available in 30-m nominal resolution at ORNL (http://dx.doi.org/10.3334/ORNLDAAC/1290). The last chapter of the book, Chapter 10, develops a method for monitoring annual vegetated land loss to urbanization with Landsat image archives. Landsat imagery has long been utilized to monitor urbanization and ecosystem change at local and regional scales. However, only a few studies have used Landsat time series to monitor urbanization at higher temporal frequencies, especially for applications focusing on large geographic areas, mainly due to the lack of efficient algorithms and computation facilities to handle large data volume (Zhang and Weng, 2016, 2017). The developed method first generates annual Landsat cloud/shadow-free NDVI mosaics and then NDVI time series spanning the period from 2000 to 2010. Next, change and stable models to identify change time points in the time series were applied. Finally, annual vegetated land loss to urbanization was extracted. The proposed method was experimentally applied in Shanghai, China, and implemented on the Google Earth Engine. The result shows annual ecosystem disturbance caused by urban expansion was well identified, with a change detection accuracy of over 80%.

References

Gao, F., de Colstoun, E.B., Ma, R., Weng, Q., Masek, J.G., Chen, J., Pan, Y. and Song, C. 2012. Mapping impervious surface expansion using medium-resolution satellite image time series: A case study in the Yangtze River Delta, China. *International Journal of Remote Sensing*, 33(24), 7609–7628.

Gutman, G., Byrnes, R., Masek, J., Covington, S., Justice, C., Franks, S. and Headley, R. 2008. Towards monitoring lad-cover and land-use changes at a global scale: The global land survey 2005. *Photogrammetric Engineering & Remote Sensing*, 74(1), 6–10.

Sexton, J.O., Song, X.-P., Huang, C., Channan, S., Baker, M.E. and Townshend, J.R. 2013. Urban growth of the Washington, D.C.–Baltimore, MD metropolitan region from 1984 to 2010 by annual, Landsat-based estimates of impervious cover. *Remote Sensing of Environment*, 129, 42–53.

Weng, Q. 2012. *An Introduction to Contemporary Remote Sensing*. New York: McGraw-Hill Professional, p. 320.

Weng, Q. and Fu, P. 2014. Modeling annual parameters of land surface temperature variations and evaluating the impact of cloud cover using time series of Landsat TIR data. *Remote Sensing of Environment*, 140, 267–278.

Weng, Q., Fu, P. and Gao, F. 2014. Generating daily land surface temperature at Landsat resolution by fusing Landsat and MODIS data. *Remote Sensing of Environment*, 145, 55–67. DOI: 10.1016/j.rse.2014.02.003.

Woodcock, C.E., Allen, R., Anderson, M., Belward, A., Bindschadler, R., Cohen, W., Gao, F. et al. 2008. Free Access to Landsat imagery. *Science*, 320, 1011–1011.

Zhang, L. and Weng, Q. 2016. Annual dynamics of impervious surface in the Pearl River Delta, China, from 1988 to 2013, using time series Landsat data. *ISPRS Journal of Photogrammetry and Remote Sensing*, 113(3), 86–96.

Zhu, Z. and Woodcock, C.E. 2014. Automated cloud, cloud shadow, and snow detection in multitemporal Landsat data: An algorithm designed specifically for monitoring land cover change. *Remote Sensing of Environment*, 152, 217–234.

Acknowledgments

Let me first thank all the contributors for making this endeavor possible. Furthermore, I offer my deepest appreciation to all the reviewers, who have taken precious time from their busy schedules to review the chapters submitted to this book. Finally, I appreciate my family for their love and support. It is my hope that the publication of this book will provide strong stimulation to students, researchers, and practitioners to conduct more in-depth studies on time series image processing, and will open up new opportunities for years ahead in the remote sensing and image science community.

The reviewers of this book are listed below in alphabetical order:

- Elfatih Abdel-Rahman
- Gang Chen
- Chengbin Deng
- Chunyun Diao
- Peng Fu
- Song Guo
- Clément Mallet
- Zina Mitraka
- Abel Ramoelo
- Miaogen Shen
- James Voogt
- Zhuosen Wang
- Yanhua Xie

Editor

Qihao Weng is the Director of the Center for Urban and Environmental Change and a Professor at Indiana State University and worked as a Senior Fellow at the National Aeronautics and Space Administration from December 2008 to December 2009. He received his PhD from the University of Georgia in 1999. Weng is currently the Lead of Group on Earth Observation (GEO) Global Urban Observation and Information Initiative, and serves as an Editor-in-Chief of *ISPRS Journal of Photogrammetry and Remote Sensing* and the Series Editor of *Taylor & Francis Series in Remote Sensing Applications*. He has been the Organizer and Program Committee Chair of the biennial IEEE/ISPRS/GEO sponsored International Workshop on Earth Observation and Remote Sensing Applications conference series since 2008, a National Director of American Society for Photogrammetry and Remote Sensing from 2007 to 2010, and a panelist of U.S. DOE's Cool Roofs Roadmap and Strategy in 2010. In 2008, Weng received a prestigious NASA senior fellowship. He is also the recipient of the Outstanding Contributions Award in Remote Sensing in 2011 and the Willard and Ruby S. Miller Award in 2015 for his outstanding contributions to geography, both from the American Association of Geographers. In 2005 at Indiana State University, he was selected as a Lilly Foundation Faculty Fellow and in the following year, he also received the Theodore Dreiser Distinguished Research Award. In addition, he was the recipient of 2010 Erdas Award for Best Scientific Paper in Remote Sensing (1st place) and 1999 Robert E. Altenhofen Memorial Scholarship Award, which were both awarded by American Society for Photogrammetry and Remote Sensing. He was also awarded the Best Student-Authored Paper Award by International Geographic Information Foundation in 1998. Weng has been invited to give 100 talks to organizations and conferences held in the US, Canada, China, Brazil, Greece, UAE, and Hong Kong. Weng's research focuses on remote sensing applications to urban environmental and ecological systems, land-use and land-cover changes, urbanization impacts, environmental modeling, and human-environment interactions. Weng is the author of 210 articles and 10 books. According to Google Scholar, as of October 2017, his SCI citation reached 12,227 (H-index of 50), and 28 of his publications had more than 100 citations each. Weng's research has been supported by funding agencies that include NSF, NASA, USGS, USAID, NOAA, National Geographic Society, European Space Agency, and Indiana Department of Natural Resources.

Contributors

Benjamin Bechtel
Center for Earth System Research
    and Sustainability
Universität Hamburg
Hamburg, Germany

Shanshan Cai
Department of Geography
The Ohio State University
Columbus, Ohio

Jin Chen
Institute of Remote Sensing Science
    and Engineering
Beijing Normal University
Beijing, China

Chengbin Deng
Department of Geography
State University of New York
    at Binghamton
Binghamton, New York

Iryna Dronova
Department of Landscape
    Architecture and Environmental
    Planning
University of California, Berkeley
Berkeley, California

Andrew Haywood
Cooperative Research Centre for
    Spatial Information (CRCSI)
Carlton, Victoria, Australia

and

European Forest Institute
Sant Leopold Pavilion
Barcelona, Spain

Binbin He
School of Resources and
    Environment
and
Center for Information Geoscience
University of Electronic Science
    and Technology of China
Chengdu, Sichuan, China

Eileen H. Helmer
International Institute of Tropical
    Forestry
USDA Forest Service
Río Piedras, Puerto Rico

Samuel Hislop
Geospatial Sciences
School of Science
RMIT University
Melbourne, Victoria, Australia

and

Cooperative Research Centre for
    Spatial Information (CRCSI)
Carlton, Victoria, Australia

and

Faculty of Geo-Information
    Science and Earth Observation
    (ITC)
University of Twente
Enschede, The Netherlands

Chengquan Huang
Department of Geographical Sciences
University of Maryland
College Park, Maryland

Simon Jones
Geospatial Sciences
School of Science
RMIT University
Melbourne, Victoria, Australia

and

Cooperative Research Centre for
    Spatial Information (CRCSI)
Carlton, Victoria, Australia

Zhang Lei
School of Geodesy and Geomatics
Wuhan University
Wuhan, China

and

Center for Urban and
    Environmental Change
Department of Earth and
    Environmental Systems
Indiana State University
Terre Haute, Indiana

Lu Liang
Arkansas Forest Resources
    Center
University of Arkansas Division
    of Agriculture
and
School of Forestry and Natural
    Resources
University of Arkansas at
    Monticello
Monticello, Arkansas

Desheng Liu
Department of Geography
The Ohio State University
Columbus, Ohio

Trung H. Nguyen
Geospatial Sciences
School of Science
RMIT University
Melbourne, Victoria, Australia

and

Cooperative Research Centre for
    Spatial Information (CRCSI)
Carlton, Victoria, Australia

Bhartendu Pandey
Yale School of Forestry and
    Environmental Studies
Yale University
New Haven, Connecticut

Shi Qiu
Department of Geosciences
Texas Tech University
Lubbock, Texas

and

School of Resources and Environment
University of Electronic Science
    and Technology of China
Chengdu, Sichuan, China

Yuhan Rao
Department of Geographical
    Sciences
University of Maryland
College Park, Maryland

Panagiotis Sismanidis
Institute for Astronomy, Astro-
    physics, Space Applications and
    Remote Sensing
National Observatory of Athens
Athens, Greece

Mariela Soto-Berelov
Geospatial Sciences
School of Science
RMIT University
Melbourne, Victoria, Australia

and

Cooperative Research Centre for
  Spatial Information (CRCSI)
Carlton, Victoria, Australia

Qihao Weng
Center for Urban and
  Environmental Change
Department of Earth and
  Environmental Systems
Indiana State University
Terre Haute, Indiana

Qingling Zhang
School of Aeronautics and
  Astronautics
Sun Yat-Sen University
Shenzhen, Guangdong, China

Feng Zhao
Department of Geographical
  Sciences
University of Maryland
College Park, Maryland

Xiaolin Zhu
Department of Land Surveying
  and Geo-Informatics
The Hong Kong Polytechnic
  University
Hong Kong, China

Zhe Zhu
Department of Geosciences
and
Center for Geospatial Technology
and
Climate Science Center
Texas Tech University
Lubbock, Texas

Part I

Time Series Image/Data Generation

1

Cloud and Cloud Shadow Detection for Landsat Images: The Fundamental Basis for Analyzing Landsat Time Series

Zhe Zhu, Shi Qiu, Binbin He, and Chengbin Deng

CONTENTS

Brief Summary..4
1.1   Introduction...4
1.2   Landsat Data and Reference Masks 5
     1.2.1   Landsat Data...5
     1.2.2   Manual Masks of Landsat Cloud and Cloud Shadow................7
1.3   Cloud and Cloud Shadow Detection Based on a Single-Date Landsat Image 8
     1.3.1   Physical-Rules-Based Cloud and Cloud Shadow Detection........8
           1.3.1.1   Physical-Rules-Based Cloud Detection Algorithms......8
           1.3.1.2   Physical-Rules-Based Cloud Shadow Detection Algorithms...12
     1.3.2   Machine-Learning-Based Cloud and Cloud Shadow Detection 14
1.4   Cloud and Cloud Shadow Detection Based on Multitemporal Landsat Images 14
     1.4.1   Cloud Detection Based on Multitemporal Landsat Images 15
     1.4.2   Cloud Shadow Detection Based on Multitemporal Landsat Images...16
1.5   Discussions 17
     1.5.1   Comparison of Different Algorithms 17
     1.5.2   Challenges..17
     1.5.3   Future Development..18
           1.5.3.1   Spatial Information..18
           1.5.3.2   Temporal Frequency...18
           1.5.3.3   Haze/Thin Cloud Removal...18
1.6   Conclusion 19
References 19

Brief Summary

Cloud and cloud shadow detection is the fundamental basis for analyzing Landsat time series. This chapter provides a comprehensive review of all the cloud and cloud shadow detection algorithms designed explicitly for Landsat images. This review provides guidance on the selection of cloud and cloud shadow detection algorithms for various applications using Landsat time series.

1.1 Introduction

Landsat satellites have been widely used for a variety of remote sensing applications, such as change detection (Collins and Woodcock, 1996; Xian et al., 2009), land cover classification (Homer et al., 2004; Yuan et al., 2005), biomass estimation (Zheng et al., 2004; Lu, 2005), and leaf area index retrieval (Chen and Cihlar, 1996; Fassnacht et al., 1997). Nevertheless, for decades, most of the analyses were based on a single or a few cloud free Landsat images acquired at different dates, due to the high cost of Landsat images prior to 2008 (Loveland and Dwyer, 2012). Free and open access to the entire Landsat archive in 2008 has changed the story entirely (Woodcock et al., 2008; Wulder et al., 2012). Landsat data are being downloaded for an unprecedented variety of applications. Many of them require frequent Landsat observations for the same location – Landsat Time Series (LTS). The Landsat Global Archive Consolidation (LGAC) initiative has added 3.2 million Landsat images to the U.S. Geological Survey (USGS) Earth Resources Observation and Science (EROS) Center (Wulder et al., 2016), which has made time series analysis with LTS even more popular. Decreasing data storage costs and increasing computing power have further stimulated the use of LTS.

Though time series analysis based on LTS has attracted much attention, automated cloud and cloud shadow detection has been and remains a major obstacle. The presence of clouds and cloud shadows reduces the usability of the Landsat image which makes it difficult for any kind of remote sensing applications. For coarse resolution images, such as from the Advanced Very High Resolution Radiometer (AVHRR) and Moderate Resolution Imaging Spectroradiometer (MODIS), there are many mature operational algorithms for detecting clouds and cloud shadows (Derrien et al., 1993; Ackerman et al., 1998). However, for moderate resolution satellites, like Landsat, there were no algorithms that could provide cloud and cloud shadow masks at the pixel level. This is not surprising because Landsat images were not affordable, each of which previously cost more than 400 U.S. dollars per image. Even when cloudy Landsat images are used, most of the time only a small number of images are needed, and manual interpretation of clouds and their shadows in the images is feasible. However, when these financial constraints were lifted (Woodcock et al., 2008), an unprecedented demand arose for automatically processing a massive number of Landsat images for time series analysis. Manual interpretation of cloud and cloud shadow was no longer acceptable.

1.2 Landsat Data and Reference Masks

1.2.1 Landsat Data

Since 1972, Landsat satellites have provided a continuous Earth observation data record. Landsats 1–5 carried the Multispectral Scanner System (MSS) sensor with 60-meter spatial resolution. The MSS only collected images with four spectral bands, including green, red, and two Near InfraRed (NIR) bands (Table 1.1). Note that the Landsat 3 MSS also included a Thermal Infrared (TIR) band, but failed shortly after launch. The fewer bands result in known difficulties in detecting clouds and cloud shadows (Braaten et al., 2015). However, the MSS images are still crucial for LTS related analyses (Pflugmacher et al., 2012). Since the launch of Landsat 4 in 1982, the Thematic Mapper (TM) has provided more spectral information at 30-meter spatial resolution (Table 1.1). The TM sensor was also carried on Landsat 5, which was launched on March 1, 1984, and functioned for over 28 years until 2012. Landsat 7, carrying the Enhanced Thematic Mapper Plus (ETM+), was launched on April 15, 1999 (Table 1.1). This instrument also has a 30-meter spatial resolution and improved radiometric and geometric calibration accuracies, but the Scan Line Corrector (SLC) has failed since May 31, 2003. Both TM and ETM+ have a TIR band at a spatial resolution of 120-meter and 60-meter, respectively. Landsat 8 was launched on February 11, 2013. It has two sensors: Operational Land Imager (OLI) and Thermal Infrared Sensor (TIRS) (Table 1.1). The OLI instrument provides 30-meter resolution optical data, while TIRS provides 100-meter resolution TIR data. Note that the TIRS has a shorter design life compared to the OLI. Additionally, the new OLI added the new blue band (Band 1: 0.435–0.451 μm) and the cirrus band (Band 9: 1.363–1.384 μm) with 30-meter spatial resolution.

Although each Landsat satellite can cover global land every 16 days, many of the observations are inevitably impacted by clouds and cloud shadows. Figure 1.1 illustrates mean global cloud cover calculated based on all available Landsat 8 daytime images acquired between September 2013 and August 2017. The cloud cover information for each Landsat Path/Row is calculated based on the metadata of Landsat 8 images downloaded from the USGS Landsat Bulk Metadata Service (https://landsat.usgs.gov/landsat-bulk-metadata-service), which is derived based on an algorithm called Fmask

TABLE 1.1

Landsat 1–5 MSS, Landsat 4–5 TM, Landsat 7 ETM+ and Landsat 8 OLI Sensor Characteristics

Landsat 1–5 MSS Bands (μm)	Landsat 4–5 TM Bands (μm)	Landsat 7 ETM+ Bands (μm)	Landsat 8 OLI/TIRS Bands (μm)
	Band 1 (0.45–0.52)	Band 1 (0.45–0.52)	Band 1 (0.435–0.451)
Band 4 (0.50–0.60)	Band 2 (0.52–0.60)	Band 2 (0.52–0.60)	Band 2 (0.452–0.512)
Band 5 (0.60–0.70)	Band 3 (0.63–0.69)	Band 3 (0.63–0.69)	Band 3 (0.533–0.590)
Band 6 (0.70–0.80)	Band 4 (0.76–0.90)	Band 4 (0.77–0.90)	Band 4 (0.636–0.673)
Band 7 (0.80–1.10)			
	Band 5 (1.55–1.75)	Band 5 (1.55–1.75)	Band 5 (0.851–0.879)
Band 8 (10.40–12.50) Landsat 3 only[a]	Band 6 (10.40–12.50)	Band 6 (10.40–12.50)	Band 6 (1.566–1.651)
	Band 7 (2.08–2.35)	Band 7 (2.09–2.35)	Band 7 (2.107–2.294)
		Band 8 (0.52–0.90)	Band 8 (0.503–0.676)
			Band 9 (1.363–1.384)
			Band 10 (10.60–11.19)
			Band 11 (11.50–12.51)

[a] Indicates that the thermal band of the Landsat 3 MSS was unsuccessful and not available.

(Zhu and Woodcock, 2012; Zhu et al., 2015). Extremely high cloud cover is observed in tropical rainforest regions, while for arid places, such as desert or dryland regions, cloud cover is relatively low. The mean global cloud cover contained in the Landsat images is approximately 41.59%, which means that clouds impact almost half of the Landsat observations.

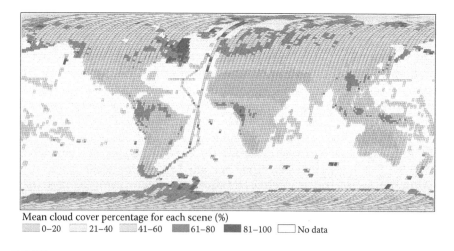

Mean cloud cover percentage for each scene (%)
▨ 0–20   ▨ 21–40   ▨ 41–60   ▨ 61–80   ▨ 81–100   ☐ No data

FIGURE 1.1

(See color insert.) Mean global cloud cover percentage calculated based on all available Landsat 8 images acquired between September 2013 and August 2017. A total of 966,708 Landsat 8 images are used. The mean global cloud cover percentage from all Landsat 8 observations is 41.59%.

1.2.2 Manual Masks of Landsat Cloud and Cloud Shadow

Manually interpreted cloud and cloud shadow masks are the most important data source for developing and/or validating the cloud and cloud shadow detection algorithms (Irish et al., 2006; Zhu and Woodcock, 2012; Hughes and Hayes, 2014; Foga et al., 2017; Qiu et al., 2017). At present, there are three publicly available, manually interpreted cloud and cloud shadow masks derived from Landsat images (Table 1.2), including "L7 Irish" masks for Landsat 7 data (USGS, 2016a), "L8 SPARCS" masks for Landsat 8 data (USGS, 2016b), and "L8 Biome" masks for Landsat 8 data (USGS, 2016c). These masks are manually interpreted based on Landsat images randomly selected from different locations, which cover a variety of land cover types, and the cloud cover percentage within each manual mask also varies substantially. The "L7 Irish" manual masks were first created to systematically cover the global environments and different cloud conditions (Irish et al., 2006). The "L7 Irish" masks were produced based on Landsat 7 ETM+ images by visual interpretation of full resolution images with different Landsat band combinations, and their average error was estimated at approximately 7% (Oreopoulos et al., 2011). The "L8 SPARCS" manual masks were created manually from Landsat 8 OLI images by Hughes and Hayes (2014), which was used to validate Spatial Procedures for Automated Removal of Cloud and Shadow (SPARCS) algorithm. Note that those manual cloud and cloud shadow masks are provided at a 3 km by 3 km Landsat subset (1000×1000 30-meter pixels), with around 4% of pixels being ambiguous (Foga et al., 2017). The manual cloud and cloud shadow masks in "L8 Biome" are designed for Landsat 8 OLI/TIRS images, which were randomly selected from different locations around the world using a biome-based stratified sampling approach. Their corresponding cloud and cloud shadow masks were produced by multiple visual criteria (such as brightness, shape, and texture) with various band combinations by a single analyst (Foga et al., 2017). This new dataset achieved better accuracy than the "L7 Irish," due to the multiple visual criteria it used (Foga et al., 2017).

TABLE 1.2

Manual Cloud and Cloud Shadow Masks Derived from Landsat Images

Name	Sensor	Number of Images	Date Range		Error	Reference
			Start	End		
L7 Irish	ETM+ (SLC on)	206 (45)	06/06/2000	12/30/2001	7.00%	USGS (2016a)
L8 SPARCS	OLI	80 (80)	05/12/2013	11/02/2014	4.00%	USGS (2016b)
L8 Biome	OLI	96 (33)	04/13/2013	11/05/2014	Less than 7.00%	USGS (2016c)

Note that the all images contain manual cloud masks. The numbers in the brackets indicate the number of cloud shadow masks for each dataset.

1.3 Cloud and Cloud Shadow Detection Based on a Single-Date Landsat Image

Recently, many cloud and cloud shadow detection algorithms have been developed for Landsat images (Table 1.3). Among them, some were proposed by using a single-date Landsat image (hereafter single-date algorithms), and we can classify these single-date algorithms into two categories: physical-rules-based and machine-learning-based algorithms (Table 1.3).

1.3.1 Physical-Rules-Based Cloud and Cloud Shadow Detection

1.3.1.1 Physical-Rules-Based Cloud Detection Algorithms

The physical-rules-based algorithms detect clouds by identifying their physical characteristics of clouds, that are "bright", "white", "cold", and "high" (Irish, 2000; Zhu et al., 2015). Compared to other land cover types, the reflectance of cloud is much higher in almost all wavelengths, which makes clouds look "bright". Therefore, we can use some simple thresholds in the spectral bands to exclude clear sky pixels that are not bright enough. Clouds are "white" due to the similar reflectance in all wavelengths, particularly in the visible bands. In this case, some indices such as "whiteness" (Zhu and Woodcock, 2012), Normalized Difference Vegetation Index (NDVI), and Normalized Difference Snow Index (NDSI) can be used to separate clouds from clear sky pixels that are not white enough. Moreover, clouds are "cold" because they are usually high in the air, and the temperature of clouds follows the environmental lapse rate—the higher the clouds, the colder the temperature. This characteristic can be successfully captured by the thermal band from Landsat TM, ETM+, and TIRS instruments, which can further separate clouds from similar bright and white land surfaces (e.g., barren sand, soil, rock, snow/ice, etc.). Additionally, as clouds are usually "high" in the sky, the path for water vapor over clouds is much shorter than that for places without clouds. Therefore, the water vapor absorption band (or the cirrus band) is especially helpful in identifying higher altitude clouds.

Most of the algorithms are developed for Landsat TM and ETM+ images. Historically, the Automated Cloud Cover Assessment (ACCA) was used to provide cloud cover percentage in Landsat TM and ETM+ images (Irish, 2000; Irish et al., 2006). With several spectral filters, ACCA works well for estimating a cloud cover score for each image but is not sufficiently precise in identifying the locations and boundaries of clouds (Zhu and Woodcock, 2012). Besides, ACCA fails to identify warm cirrus clouds and may misidentify snow/ice as clouds, mainly because the static thresholds in ACCA are insufficient to capture the various kinds of clouds and the variety of land surface types. To better distinguish cloud from snow/ice, Choi and Bindschadler (2004) used the cloud and cloud shadow geometry matching approach iteratively

TABLE 1.3

Characteristics of Different Cloud and Cloud Shadow Detection Algorithms for Landsat Images

Category	Algorithm Name	Landsat Sensor	Cloud/Shadow	Ancillary Data	Overall Accuracy	Reference
Single-date: Physical rules	MFmask	TM ETM+ OLI/TIRS	Both	DEM	96%	Qiu et al. (2017)
	LSR 8	OLI/TIRS	Both	N/A	N/A	Vermote et al. (2016)
	UDTCDA	OLI/TIRS	Cloud	MOD09A1	N/A	Sun et al. (2016)
	MSScvm	MSS	Both	DEM	84%	Braaten et al. (2015)
	ELTK	OLI/TIRS	Cloud	N/A	N/A	Wilson and Oreopoulos (2013)
	Fmask	TM ETM+ OLI/TIRS	Both	N/A	96%	Zhu and Woodcock (2012), Zhu et al. (2015)
	N/A	TM ETM+	Both	DEM	88%~99%	Huang et al. (2010)
	LTK	TM ETM+ OLI/TIRS	Cloud	N/A	93%	Oreopoulos et al. (2011)
	LEDAPS	TM ETM+	Both	Air temperature from NCEP	N/A	Vermote and Saleous (2007)
	CDSM/ANTD	ETM+	Both	N/A	N/A	Choi and Bindschadler (2004)
	ACCA	TM ETM+ OLI/TIRS	Cloud	N/A	N/A	Irish (2000), Irish et al. (2006)
Single-date: Machine learning	N/A	OLI	Cloud	N/A	N/A	Zhou et al. (2016)
	SPARCS	ETM+	Both	N/A	99%	Hughes and Hayes (2014)
	See5	OLI	Cloud	N/A	89%	Scaramuzza et al. (2012)
	AT-ACCA	OLI	Cloud	N/A	90%	Scaramuzza et al. (2012)
	N/A	ETM+	Both	N/A	N/A	Potapov et al. (2011)
	N/A	ETM+	Cloud	N/A	N/A	Roy et al. (2010)
	N/A	MSS	Cloud	N/A	93%	Lee et al. (1990)

(Continued)

TABLE 1.3 (*Continued*)

Characteristics of Different Cloud and Cloud Shadow Detection Algorithms for Landsat Images

Category	Algorithm Name	Landsat Sensor	Cloud/ Shadow	Ancillary Data	Overall Accuracy	Reference
Multi-date	IHOT	MSS TM ETM+	Cloud	N/A	N/A	Chen et al. (2015)
	Tmask[a]	TM ETM+ OLI/TIRS	Both	N/A	N/A	Zhu and Woodcock (2014)
	N/A[a]	TM ETM+	Both	N/A	97%	Goodwin et al. (2013)
	N/A	TM ETM+	Both	N/A	N/A	Jin et al. (2013)
	MTCD[a]	TM ETM+	Both	Sentinel-2 data	N/A	Hagolle et al. (2010)
	N/A	TM	Both	N/A	N/A	Wang et al. (1999)

Note: MFmask: Mountainous Fmask; LSR 8: Landsat 8 Surface Reflectance product; UDTCDA: Universal Dynamic Threshold Cloud Detection Algorithm; MSScvm: MSS clear-view-mask; ELTK: Enhanced LTK; Fmask: Function of mask; LTK: Luo Trishchenko Khlopenkov; LEDAPS: Landsat Ecosystem Disturbance Adaptive Processing System; CDSM/ANTD: Cloud Detection using Shadow Matching/Automatic NDSI Threshold Decision; ACCA: Automatic Cloud Cover Assessment; SPARCS: Spatial Procedures for Automated Removal of Cloud and Shadow; See5: C5.0 algorithm used to generate a decision tree. AT-ACCA: Artificial Thermal-Automated Cloud Cover Algorithm; IHOT: Iterative Haze Optimized Transformation; Tmask: multiTemporal mask; MTCD: Multi-Temporal Cloud Detection; TM: Thematic Mapper; ETM+: Enhanced Thematic Mapper Plus; OLI: Operational Land Imager; TIRS: Thermal Infrared Sensor; DEM: Digital Elevation Model; MOD09A1: MODerate resolution Imaging Spectroradiometer surface reflectance product; NCEP: National Centers for Environmental Prediction.

[a]   Indicates the algorithms based on Landsat time series.

to determine the optimal threshold of NDSI for each Landsat image in cloud detection. This approach works well over ice sheets, but it is time-consuming and only works on the surface of ice sheets. Vermote and Saleous (2007) proposed a cloud detection algorithm for Landsat TM and ETM+ images, and the detection results are provided as one of the internal products in the Landsat Ecosystem Disturbance Adaptive Processing System (LEDAPS) atmosphere correction software. This algorithm needs surface temperature from the National Centers for Environmental Prediction (NCEP) as ancillary data to generate a surface temperature reference layer for cloud detection. Huang et al. (2010) constructed a spectral temperature space to identify clouds in Landsat image using clear sky forest pixels as a reference. This method works well over forest areas but has not been fully tested for non-forest areas. By revisiting the Luo Trishchenko Khlopenkov (LTK) scene identification algorithm initially developed for the MODIS image (Luo et al., 2008), Oreopoulos et al. (2011) modified this algorithm to detect clouds in Landsat 7 ETM+ data using simple thresholds derived for the blue, red, NIR, and Short-Wave Infrared (SWIR) bands (no thermal band). Recently, Zhu and Woodcock (2012) developed the Fmask (Function of mask) algorithm that detects cloud by using a scene-based threshold. This method is suitable for the Landsats 4–8 data and can generate a cloud probability layer. Users can adjust the threshold of cloud probability to determine cloud masks. The default threshold (global optimal) is 22.5%. If large omissions are found, a smaller threshold (e.g., 12.5%) is recommended, and if large commissions are observed, a higher threshold (e.g., 50%) is recommended. This method has also been successfully integrated into the Landsat surface reflectance Climate Data Record (CDR) and Collection 1 Quality Assessment (QA) band provided by the USGS Earth Resources Observation and Science (EROS) Center. In the Fmask algorithm, the thermal band is one of the important inputs, as it can capture the "cold" character of clouds (Zhu et al., 2015). However, the temperature for clear sky pixels can also vary widely due to substantial changes in elevation, and this will lead to commission and omission errors in cloud detection in mountainous areas. To reduce this issue, Qiu et al. (2017) provided a Mountainous Fmask (MFmask) algorithm that normalizes the thermal band with Digital Elevation Models (DEMs) based on a simple linear temperature-elevation model.

There are also algorithms explicitly designed for Landsat 8 images, many of which take advantage of the new blue and cirrus bands equipped in Landsat OLI. Wilson and Oreopoulos (2013) further modified the aforementioned LTK algorithm by including the cirrus band to detect cloud better. Zhu et al. (2015) also designed a cloud detection algorithm for Landsat 8 images by calculating a thin cloud probability layer from the cirrus band, and achieved better accuracy than the Fmask algorithm designed initially for TM and ETM+ images. Vermote et al. (2016) proposed a new cloud detection algorithm for Landsat 8, which used the inversion "residual" from the two blue bands and the cirrus band reflectance. To minimize the influences of cloud detection from mixed pixels, complex surface structures, and atmospheric factors, Sun et al. (2016) presented a Universal Dynamic Threshold Cloud Detection Algorithm (UDTCDA) for Landsat 8 OLI images, but only the blue, green, red, NIR, and SWIR bands were used. The dynamic threshold in this method was determined based on MODIS monthly surface reflectance database, which was established based on the long-time series of MODIS 8-day synthetic surface reflectance products.

Very few algorithms have been designed for the Landsat MSS image, due to the limited number of spectral bands within the MSS sensor. To address this issue, Braaten et al. (2015) proposed a simple and automated cloud detection algorithm relying on green band brightness and the normalized difference between the green and red bands and achieved comparable accuracies to the Fmask algorithm.

1.3.1.2 Physical-Rules-Based Cloud Shadow Detection Algorithms

Detecting cloud shadows for Landsat images is more difficult than detecting clouds due to the spectral similarity of cloud shadows and dark surfaces. Cloud shadows are usually detected based on physical rules derived from the cloud shadow geometry.

Previously, the cloud shadow detection algorithms were developed based on simple spectral tests according to the dark features. However, it is difficult to directly use thresholds to determine cloud shadows because their spectral signatures are very similar to other dark surfaces (e.g., terrain shadows, wetlands, dark urban, etc.). Fortunately, the geometry-based cloud shadow detection has shown relatively good results. The geometry-based cloud shadow detection approach is based on the projection of cloud object onto the local plane of the Earth with respect to a direction of incoming solar radiation (Berendes et al., 1992; Le Hégarat-Mascle and André, 2009; Simpson et al., 2000). The relative positions of the sun, the satellite, and the cloud can be used to predict the cloud shadow observed in the satellite images (Figure 1.2). Methods for detecting cloud shadow based on geometry can be grouped into two categories: shape-similarity-match and cloud-height-estimation.

The shape-similarity-match approach detects cloud shadow by matching cloud shadows with cloud objects, assuming that cloud and cloud shadow shape are similar (Gurney, 1982; Berendes et al., 1992). Gurney (1982) assumed that a cumulus cloud is approximated in shape by its associated shadow and matched cloud shadows with clouds. Berendes et al. (1992) developed a semi-automated methodology for estimating cumulus cloud base height using Landsat data by matching cloud edges with their corresponding shadow edges. Due to the absence of the thermal band, Braaten et al. (2015) used cloud projection to identify cloud shadow in Landsat MSS image based on their geometry information. Although the computation of cloud and cloud shadow match is time-consuming and may result in some mismatches, this approach

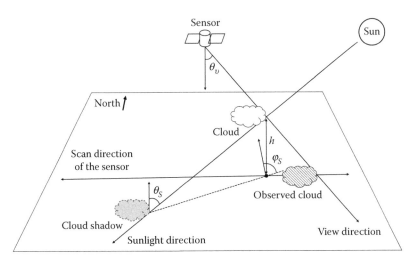

FIGURE 1.2
Sun/cloud/shadow geometry in Landsat image. Note that θ_s is the solar zenith angle, φ_s is the solar azimuth angle, θ_v is the satellite view zenith angle, and h is the cloud height. The position of the real cloud may shift from the cloud directly observed by the Landsat sensor.

is still an attractive routine, especially for images without the thermal band (e.g., Landsat MSS).

The cloud-height-estimation method uses a constant lapse rate to estimate cloud top height based on the Brightness Temperature (BT) difference between cloud top and the land surface. When the cloud height is known, the clouds can be easily projected to predict their associate shadows on the ground based on their geometry relationship (Vermote and Saleous, 2007; Huang et al., 2010). Vermote and Saleous (2007) detected cloud shadow for Landsat data using a geometric determination of shadow based on the cloud mask and the estimated altitude of cloud derived from the BT and a conversion factor range. Huang et al. (2010) identified cloud shadow based on the dark spectral features, the temperature-based cloud height estimation with a constant lapse rate, and the solar illumination geometry. Those methods can work well for thick clouds but are less ideal for the semitransparent clouds, of which the BT will be a mixture of thin cloud and the surface underneath.

Zhu and Woodcock (2012) and Zhu et al. (2015) calculated cloud shadows by combining the previous shape-similarity-matching and the cloud-height-estimation methods, and treated the cloud as a 3D object. This algorithm works well for most cases, but it may fail to detect the correct cloud shadows for places of large topographic change and terrain shadows. To address those issues, Qiu et al. (2017) applied a double-projection method to calculate cloud shadow shape and removed terrain shadows using a topographic correction model with the aid of DEMs. In addition, this improved Fmask algorithm can estimate a more accurate cloud height based on heights estimated from neighboring clouds, that also improves the detection of cloud shadow for Landsat images.

1.3.2 Machine-Learning-Based Cloud and Cloud Shadow Detection

Regarding the machine-learning-based algorithms, clouds and/or cloud shadows are generally treated as a cover type and are identified using a certain classifier trained by previously collected training dataset – supervised classification. Lee et al. (1990) combined texture-based features in a network to discriminate clouds in Landsat MSS images and achieved an overall cloud identification accuracy of 93%. Recognizing that omission and commission errors for cloud detection will always occur in large datasets for ACCA, Roy et al. (2010) implemented both the ACCA algorithm and a classification tree approach to detect clouds using a large number of training pixels from a global Landsat Level 1G database. Potapov et al. (2011) also manually selected lots of cloud pixels as training data based on 21 Landsat images from different years and different regions and built a single tree model for cloud type to classify the clouds. Due to the possible loss of the thermal band on Landsat 8, Scaramuzza et al. (2012) expanded the ACCA pass-1 algorithm without the use of the thermal band, identifying clouds for Landsat 8 through a statistical classifier C5.0 (a classification tree) based on many randomly sampled pixels from a series of training images. Hughes and Hayes (2014) also explored the inclusion of spatial information as an input to a neural network classifier on identifying and classifying clouds for Landsat images. Zhou et al. (2016) utilized the traditional threshold to obtain a coarse cloud mask and then used the Support Vector Machine (SVM) classifier to detect clouds in Landsat 8 images. Though all these investigations pointed out the usefulness of machine-learning-based methods in cloud detection, most require a certain level of knowledge of cloud or surface conditions within the images (as training data) and commonly fail to detect clouds for certain unique conditions (Huang et al. 2010). Additionally, cloud shadow in Landsat image can also be detected using the machine-learning-based methods (Potapov et al., 2011; Hughes and Hayes, 2014). This method, however, heavily relies on the training dataset and has substantial omission or commission errors (Hughes and Hayes, 2014).

1.4 Cloud and Cloud Shadow Detection Based on Multitemporal Landsat Images

In addition to the single-date algorithm, cloud and cloud shadow detection algorithms based on multitemporal Landsat images have also been developed (hereafter multitemporal algorithm). Compared to the spectral or spatial features derived from a single-date Landsat image, multitemporal Landsat images can provide extra-temporal information in cloud and cloud shadow detection, and are reported to produce better cloud and cloud shadow masks

(Table 1.3). The basic idea of these algorithms is that clouds and cloud shadows can be easily detected by comparing an observed image with a clear sky reference (image differencing), as the presence of clouds and cloud shadows will result in sudden changes of Landsat reflectance (Wang et al., 1999; Jin et al., 2013).

1.4.1 Cloud Detection Based on Multitemporal Landsat Images

For a long time, multitemporal cloud detection algorithms were only based on two-date or multi-date Landsat images. Early on, Wang et al. (1999) proposed the use of two-date Landsat TM images to find clouds by image differencing. This proposed method first coarsely finds the clouds for the two Landsat TM images by setting a histogram-derived threshold for the brightness values, and then uses another static threshold for the absolute brightness difference between the two images to further ensure reliable cloud detection. Jin et al. (2013) identified clouds by incorporating Landsat blue and thermal bands from two-date images. Based on two Landsat images that have no overlapping clouds, this method first selected the relaxed clouds by differencing the blue bands from the two images and then produced the restricted clouds by eliminating some commission pixels with relatively low spectral values in the SWIR band and low temperature in the thermal band. The thresholds used in this approach were determined by measuring spectral deviation from the mean value of the input images. These methods can accurately detect cloud for the reported images, but the thresholds may not be transferable to other images. To avoid confusion between bright surfaces and haze/cloud, Chen et al. (2015) proposed an Iterative Haze Optimized Transformation (IHOT) for improving haze/clouds detection for Landsat images with the help of a corresponding clear image. By integrating an iterative procedure of regressions into the HOT (Zhang et al., 2002), the reflectance difference between hazy and clear images, and reflectance of hazy and clear images, the land surface information can be removed. The IHOT result is derived to characterize the haze contamination on Landsat images spatially. These proposed approaches are practical and straightforward only using two-date or multi-date Landsat images but heavily dependent on the quality and availability of reference images. Besides, these approaches may not work well if extensive land cover changes occurred between the acquisition dates of the reference and cloudy images.

With free and open access to the Landsat archive, time series analysis with Landsat images became possible, providing a new way to detect clouds based on higher frequency Landsat observations. The LTS itself can be used for detecting clouds. Goodwin et al. (2013) used LTS from TM/ETM+ to detect clouds. By using the minimum and median values of the blue band as a reference, this algorithm can produce better cloud masks across Queensland compared to Fmask (Zhu and Woodcock, 2012). However, it has not yet been tested in environments with different soils, vegetation cover, and structure or areas with snow/ice cover (Goodwin et al., 2013). Specifically designed for monitoring land cover change, an algorithm called Tmask (multitemporal mask) has been developed for automated masking of cloud and snow for LTS (Zhu and Woodcock, 2014). This method estimated time series models for each pixel based on "clear-sky LTS" previously filtered by the Fmask algorithm. By using a robust fitting approach, the cloud observations that are missed by Fmask will have minimal impacts on the estimation of the time series models. By comparing the model estimates with actual Landsat observations for the green, NIR, and SWIR bands, we will be able to detect any remaining cloud observations for the entire stack of Landsat images. In addition to the Landsat images, there are also algorithms developed for Landsat-like data, such as VENµS and Sentinel-2. Hagolle et al. (2010) developed the Multi-Temporal Cloud Detection (MTCD) method that detects sudden increases of reflectance in the blue band on a pixel-by-pixel basis using time series observations and tested the linear correlation of pixel neighborhoods taken from pairs of images acquired successively. The MTCD method provides better discrimination of cloudy and clear sky pixels than the ACCA method for Landsat images. However, it requires satellite data with high revisit frequency and sequential processing of the data.

1.4.2 Cloud Shadow Detection Based on Multitemporal Landsat Images

Most cloud shadow detection algorithms using multitemporal Landsat images assume that the presence of cloud shadows will lead to darker, colder, and smoother features than the regular land surface (Irish, 2000; Le Hégarat-Mascle and André, 2009). Wang et al. (1999) presented a wavelet transform approach to detect cloud shadows for two Landsat TM images automatically. Considering that the brightness changes of the cloud shadow-obscured regions are much smoother than the regions with no shadows, the absolute wavelet coefficients corresponding to cloud shadows decrease much greater amount than those of other regions. Thus, a relative contrast difference for the added result of the wavelet transform outputs was directly used to detect cloud shadow for the two Landsat images with a static threshold. Different from this complicated approach, Jin et al. (2013) detected the cloud shadows simply by differencing the SWIR and the thermal bands from two-date Landsat images and employed the geometric relationship between clouds and their corresponding shadows to reduce false positive errors. Zhu and Woodcock (2014) also identified cloud shadows for LTS by image differencing. The reference values were predicted using a time series model for each pixel. Though there are only a few cloud shadow detection approaches using multitemporal Landsat images, these methods can provide better results than the approaches based on a single-date Landsat image, especially for shadows from thin clouds (Zhu and Woodcock, 2014).

1.5 Discussions

1.5.1 Comparison of Different Algorithms

With so many different cloud and cloud shadow detection algorithms available in the literature, it is essential to compare those approaches and to provide further guidance on the selection of algorithms for those interested in using LTS. A list of most of the automated cloud and cloud shadow detection algorithms can be found in Table 1.3. We observe that the most widely used detection algorithms are based on a single-date Landsat image, probably due to the ease of implementation. Recently, Foga et al. (2017) compared the performances of several popular algorithms using 278 unique cloud validation masks over the entire globe and found that the CFmask (Fmask algorithm programmed in C) has the best overall accuracy for Landsat data. It should be noted that the methods based on multitemporal Landsat images can provide more accurate detection of cloud and cloud shadow, which is especially important for time series analysis (e.g., forest disturbance, land cover change, etc.) (Goodwin et al., 2013; Zhu and Woodcock, 2014).

1.5.2 Challenges

Clouds are easily confused with snow/ice, especially for mountaintop snow/ice (Selkowitz and Forster, 2015). These kinds of commissions can be reduced by the NDSI threshold (Zhu and Woodcock, 2012), verification of clouds with their corresponding shadows (Choi and Bindschadler, 2004), temperature normalization (Qiu et al., 2017), or composition of temporal pixels in summer season (Selkowitz and Forster, 2015). However, it is still difficult to separate clouds from snow in some circumstances (e.g., icy clouds).

The cloud shadow detection accuracy is still relatively low. The geometry projection of cloud is a good way to detect cloud shadow, but relies heavily on the previously identified cloud masks, which have commission or omission errors and subsequently result in inaccurate cloud shadows. In addition, the cloud shadows are commonly confused with other dark features, such as wetlands, dark urban, and terrain shadows. Terrain shadows can be removed using the topographic correction model with the aid of DEMs (Jin et al., 2013; Braaten et al., 2015; Qiu et al., 2017). The misidentification of cloud shadow contributed from other dark features can also be corrected based on the contextual information from the clouds' heights estimated from neighboring clouds (Qiu et al., 2017).

The use of multitemporal Landsat images can produce better cloud and cloud shadow masks by differencing new observations with reference observations. However, this kind of approach may not work well due to the range of non-cloud related variations in reflectance, such as the illumination geometry, land surface change, geometric misregistration, and variation in radiometry or atmospheric composition (Hagolle et al., 2010; Goodwin et al., 2013; Zhu and Woodcock, 2014). Furthermore, these algorithms are computationally expensive compared to cloud and cloud shadow detection algorithms based on a single-date Landsat image.

1.5.3 Future Development

1.5.3.1 Spatial Information

When designing cloud and cloud shadow detection algorithms for Landsat images, the spectral information and the temporal information have been explored extensively, but the information contained in the spatial domain is less studied (Gurney, 1982; Martins et al., 2002). We expect that more cloud and cloud shadow detection algorithms will focus on the spatial characteristics of clouds and their shadows and provide masks at higher accuracies.

1.5.3.2 Temporal Frequency

The approaches based on multitemporal Landsat images can provide more accurate cloud and cloud shadow masks, when compared to the single-date approaches (Goodwin et al., 2013; Zhu and Woodcock, 2014). In addition to Landsat data, other Landsat-like satellites have also been launched, such as Sentinel-2A/2B. The integration of multi-source images will allow more frequent observations and further improve the detection accuracy. One major restriction of the multitemporal cloud and cloud shadow detection algorithms is that these algorithms require large amounts of data and computation time. However, this will be less an issue with the rapid development of computation technology.

1.5.3.3 Haze/Thin Cloud Removal

Compared with thick clouds, thin clouds are transparent, and images covered by thin clouds include information from both the atmospheric and the ground underneath (Li et al., 2012). This gives us the opportunity to remove the impacts of haze/thin clouds. If the satellite sensor profile and the atmospheric properties are known, haze/thin clouds' impacts can be reduced by atmospheric correction (Vermote and Saleous, 2007). However, it is difficult to acquire all the atmospheric properties (Liang et al., 2001), and atmospheric correction may fail in handling the locally concentrated thin clouds (Shen et al., 2014). Methods based on multispectral transformation, such as Tasseled Cap (TC) transformation (Richter, 1996), HOT (Zhang et al., 2002), and Advanced HOT (AHOT) (Liu et al., 2011), can remove haze/thin clouds' impacts effectively. Besides, haze/thin clouds are generally distributed in the low frequency parts of the image, which can be removed by using a low-pass filter (Shen et al., 2014), such as Wavelet Analysis (WA)

(Du et al., 2002) and Homomorphic Filter (HF) (Fan and Zhang, 2011). While many haze/thin cloud removal methods are available, there are still difficulties in automated identification of haze/thin clouds using current cloud detection algorithms. This will hamper the broad applications of haze/thin cloud removal approaches.

1.6 Conclusion

Clouds and cloud shadows are a pervasive, dynamic, and unavoidable issue in Landsat images, and their accurate detection is the fundamental basis for analyzing LTS. Many cloud and/or cloud shadow detection algorithms have been proposed in the literature. For cloud detection, most approaches are based on a single-date Landsat image, which rely on physical-rules or machine-learning techniques. With the policy of free and open Landsat data, some automated cloud detection methods were developed based on multitemporal Landsat images and can achieve better results. For cloud shadow detection, the geometry-based approach is widely used in the single-date algorithms. Meanwhile, by using multitemporal Landsat images, some researchers used the image differencing method to better identify cloud shadow. In this chapter, we reviewed many automated cloud and cloud shadow detection algorithms, which can provide guidance on the selection of algorithms for those interested in using LTS.

References

Ackerman, S. A., Strabala, K.I., Menzel, W. P., Frey, R. A., Moeller, C.C., and Gumley, L.E. 1998. Discriminating clear sky from clouds with MODIS. *Journal of Geophysical Research* 103(D24): 32141–32157. doi:10.1029/1998JD200032.

Berendes, T., Sengupta, S. K., Welch, R. M., Wielicki, B. A., and Navar, M. 1992. Cumulus cloud base height estimation from high spatial resolution Landsat data: A Hough transform approach. *IEEE Transactions on Geoscience and Remote Sensing* 30(3): 430–443. doi:10.1109/36.142921.

Braaten, J. D., Cohen, W. B., and Yang, Z. 2015. Automated cloud and cloud shadow identification in Landsat MSS imagery for temperate ecosystems. *Remote Sensing of Environment* 169: 128–138. doi:10.1016/j.rse.2015.08.006.

Chen, J. M. and Cihlar, J. 1996. Retrieving leaf area index of boreal conifer forests using Landsat TM images. *Remote Sensing of Environment* 55(2): 153–162. doi:10.1016/0034-4257(95)00195-6.

Chen, S., Chen, X., Chen, J., and Jia, P. 2015. An iterative haze optimized transformation for automatic cloud/haze detection of Landsat imagery. *IEEE Transactions on Geoscience and Remote Sensing* 54(5): 2682–2694. doi:10.1109/TGRS.2015.2504369.

Choi, H. and Bindschadler, R. 2004. Cloud detection in Landsat imagery of ice sheets using shadow matching technique and automatic normalized difference snow index threshold value decision. *Remote Sensing of Environment* 91(2): 237–242. doi:10.1016/j.rse.2004.03.007.

Collins, J. B. and Woodcock, C. E. 1996. An assessment of several linear change detection techniques for mapping forest mortality using multitemporal Landsat TM data. *Remote Sensing of Environment* 56(1): 66–77. doi:10.1016/0034-4257(95)00233-2.

Derrien, M., Farki, B., Harang, L., LeGleau, H., Noyalet, A., Pochic, D., and Sairouni, A. 1993. Automatic cloud detection applied to NOAA-11 /AVHRR imagery. *Remote Sensing of Environment* 46(3): 246–267. doi:10.1016/0034-4257(93)90046-Z.

Du, Y., Guindon, B., and Cihlar, J. 2002. Haze detection and removal in high resolution satellite image with wavelet analysis. *IEEE Transactions on Geoscience and Remote Sensing* 40(1): 210–217. doi:10.1109/36.981363.

Fan, C. N. and Zhang, F. Y. 2011. Homomorphic filtering based illumination normalization method for face recognition. *Pattern Recognition Letters* 32(10): 1468–1479. doi:10.1016/j.patrec.2011.03.023.

Fassnacht, K. S., Gower, S. T., MacKenzie, M. D., Nordheim, E. V., and Lillesand, T. M. 1997. Estimating the leaf area index of North Central Wisconsin forests using the Landsat thematic mapper. *Remote Sensing of Environment* 61(2): 229–245. doi:10.1016/S0034-4257(97)00005-9.

Foga, S., Scaramuzza, P. L., Guo, S., Zhu, Z., Dilley, R. D., Beckmann, T., Schmidt, G., Dwyer, J., Hughes, M., and Laue, B. 2017. Cloud detection algorithm comparison and validation for operational Landsat data products. *Remote Sensing of Environment* 194: 379–390. doi:10.1016/j.rse.2017.03.026.

Goodwin, N. R., Collett, L. J., Denham, R. J., Flood, N., and Tindall, D. 2013. Cloud and cloud shadow screening across Queensland, Australia: An automated method for Landsat TM/ETM+ time series. *Remote Sensing of Environment* 134: 50–65. doi:10.1016/j.rse.2013.02.019.

Gurney, C. M. 1982. The use of contextual information to detect cumulus clouds and cloud shadows in Landsat data. *International Journal of Remote Sensing* 3(1): 51–62. doi:10.1080/01431168208948379.

Hagolle, O., Huc, M., Pascual, D. V., and Dedieu, G. 2010. A multi-temporal method for cloud detection, applied to FORMOSAT-2, VENµS, LANDSAT and SENTINEL-2 images. *Remote Sensing of Environment* 114(8): 1747–1755. doi:10.1016/j. rse.2010.03.002.

Homer, C., Huang, C., Yang, L., Wylie, B., and Coan, M. 2004. Development of a 2001 national land-cover database for the United States. *Photogrammetric Engineering and Remote Sensing* 70(7): 829–840. doi:10.14358/PERS.70.7.829.

Huang, C. et al. 2010. Automated masking of cloud and cloud shadow for forest change analysis using Landsat images. *International Journal of Remote Sensing* 31(20): 5449–5464. doi:10.1080/01431160903369642.

Hughes, M. J. and Hayes, D. J. 2014. Automated detection of cloud and cloud shadow in single-date Landsat imagery using neural networks and spatial post-processing. *Remote Sensing of Environment* 6(6): 4907–4926. doi:10.3390/rs6064907.

Irish, R. R. 2000. Landsat 7 automatic cloud cover assessment. *Proceedings Volume 4049, Algorithms for Multispectral, Hyperspectral, and Ultraspectral Imagery*, Orlando, FL, United States, *AeroSense: International Society for Optics and Photonics.* doi:10.1117/12.410358.

Irish, R. R., Barker, J. L., Goward, S. N., and Arvidson, T. 2006. Characterization of the Landsat-7 ETM+ Automated Cloud-Cover Assessment (ACCA) algorithm. *Photogrammetric Engineering and Remote Sensing* 72(10): 1179–1188. doi:10.14358/PERS.72.10.1179.

Jin, S., Homer, C., Yang, L., Xian, G., Fry, J., Danielson, P., and Townsend, P. A. 2013. Automated cloud and shadow detection and filling using two-date Landsat imagery in the USA. *International Journal of Remote Sensing* 34(5): 1540–1560. doi:10.1080/01431161.2012.720045.

Lee, J., Weger, R. C., Sengupta, S. K., and Welch, R. M. 1990. A neural network approach to cloud classification. *IEEE Transactions on Geoscience and Remote Sensing* 28(5): 846–855. doi:10.1109/36.58972.

Le Hégarat-Mascle, S. and André, C. 2009. Use of Markov random fields for automatic cloud/shadow detection on high resolution optical images. *ISPRS Journal of Photogrammetry and Remote Sensing* 64(4): 351–366. doi:10.1016/j.isprsjprs.2008.12.007.

Li, Q., Lu, W., Yang, J., and Wang, J. Z. 2012. Thin cloud detection of all-sky images using Markov random fields. *IEEE Geoscience and Remote Sensing Letters* 9(3): 417–421. doi:10.1109/LGRS.2011.2170953.

Liang, S., Fang, H., and Chen, M. 2001. Atmospheric correction of Landsat ETM+ land surface imagery. I. Methods. *IEEE Transactions on Geoscience and Remote Sensing* 39(11): 2490–2498. doi:10.1109/36.964986.

Liu, C. B., Hu, J. B., Lin, Y., Wu, S. H., and Huang, W. 2011. Haze detection, perfection and removal for high spatial resolution satellite imagery. *International Journal of Remote Sensing* 32(23): 8685–8697. doi:10.1080/01431161.2010.547884.

Loveland, T. R. and Dwyer, J. L. 2012. Landsat: Building a strong future. *Remote Sensing of Environment* 122: 22–29. doi:10.1016/j.rse.2011.09.022.

Lu, D. 2005. Aboveground biomass estimation using Landsat TM data in the Brazilian Amazon. *International Journal of Remote Sensing* 26(12): 2509–2525. doi:10.1080/01431160500142145.

Luo, Y., Trishchenko, A. P., and Khlopenkov, K. V. 2008. Developing clear-sky, cloud and cloud shadow mask for producing clear-sky composites at 250-meter spatial resolution for the seven MODIS land bands over Canada and North America. *Remote Sensing of Environment* 112(12): 4167–4185. doi:10.1016/j.rse.2008.06.010.

Martins, J. V., Tanré, D., Remer, L., Kaufman, Y., Mattoo, S., and Levy, R. 2002. MODIS cloud screening for remote sensing of aerosols over oceans using spatial variability. *Geophysical Research Letters* 29(12): MOD4-1–MOD4-4. doi:10.1029/2001GL013252.

Oreopoulos, L., Wilson, M. J., and Várnai, T. 2011. Implementation on Landsat data of a simple cloud-mask algorithm developed for MODIS land bands. *IEEE Geoscience and Remote Sensing Letters* 8(4): 597–601. doi:10.1109/LGRS.2010.2095409.

Pflugmacher, D., Cohen, W. B., and Kennedy, R. E. 2012. Using Landsat-derived disturbance history (1972–2010) to predict current forest structure. *Remote Sensing of Environment* 122: 146–165. doi:10.1016/j.rse.2011.09.025.

Potapov, P., Turubanova, S., and Hansen, M. C. 2011. Regional-scale boreal forest cover and change mapping using Landsat data composites for European Russia. *Remote Sensing of Environment* 115(2): 548–561. doi:10.1016/j.rse.2010.10.001.

Qiu, S., He, B., Zhu, Z., Liao Z., and Quan, X. 2017. Improving Fmask cloud and cloud shadow detection in mountainous area for Landsats 4–8 images. *Remote Sensing of Environment* 199: 107–119. doi:10.1016/j.rse.2017.07.002.

Richter, R. 1996. A spatially adaptive fast atmospheric correction algorithm. *International Journal of Remote Sensing* 17(6): 1201–1214. doi:10.1080/01431169608949077.

Roy, D. P., Ju, J., Kline, K., Scaramuzza, P. L., Kovalskyy, V., Hansen, M., Loveland, T. R., Vermote, E., and Zhang, C. 2010. Web-enabled Landsat Data (WELD): Landsat ETM+ composited mosaics of the conterminous United States. *Remote Sensing of Environment* 114(1): 35–49. doi:10.1016/j.rse.2009.08.011.

Scaramuzza, P. L., Bouchard, M. A., and Dwyer, J. L. 2012. Development of the Landsat data continuity mission cloud-cover assessment algorithms. *IEEE Transactions on Geoscience and Remote Sensing* 50(4): 1140–1154. doi:10.1109/TGRS.2011.2164087.

Selkowitz, D. J. and Forster, R. R. 2015. An automated approach for mapping persistent ice and snow cover over high latitude regions. *Remote Sensing* 8(1): 16. doi:10.3390/rs8010016.

Shen, H., Li, H., Qian, Y., Zhang, L., and Yuan, Q. 2014. An effective thin cloud removal procedure for visible remote sensing images. *ISPRS Journal of Photogrammetry and Remote Sensing* 96: 224–235. doi:10.1016/j.isprsjprs.2014.06.011.

Simpson, J. J., Jin, Z., and Stitt, J. R. 2000. Cloud shadow detection under arbitrary viewing and illumination conditions. *IEEE Transactions on Geoscience and Remote Sensing* 38(2): 972–976. doi:10.1109/36.841979.

Sun, L. et al. 2016. A universal dynamic threshold cloud detection algorithm (UDTCDA) supported by a prior surface reflectance database. *Journal of Geophysical Research: Atmospheres* 121(12): 7172–7196. doi:10.1002/2015JD024722.

U.S. Geological Survey. 2016a. L7 Irish Cloud Validation Masks. *U.S. Geological Survey data release.* doi:10.5066/F7XD0ZWC. (accessed September 26, 2017)

U.S. Geological Survey. 2016b. L8 SPARCS Cloud Validation Masks. *U.S. Geological Survey data release.* doi:10.5066/F7FB5146. (accessed September 26, 2017)

U.S. Geological Survey. 2016c. L8 Biome Cloud Validation Masks. *U.S. Geological Survey data release.* doi:10.5066/F7251GDH. (accessed September 26, 2017)

Vermote, E., Justice, C., Claverie, M., and Franch, B. 2016. Preliminary analysis of the performance of the Landsat 8/OLI land surface reflectance product. *Remote Sensing of Environment* 185: 46–56. doi:10.1016/j.rse.2016.04.008.

Vermote, E. and Saleous, N. 2007. *LEDAPS Surface Reflectance Product Description.* College Park: University of Maryland. doi:null

Wang, B., Ono, A., Muramatsu, K., and Fujiwara, N. 1999. Automated detection and removal of clouds and their shadows from Landsat TM images. *IEICE Transactions on Information and Systems* 82(2): 453–460. doi:null

Wilson, M. J. and Oreopoulos, L. 2013. Enhancing a simple MODIS CLOUD mask algorithm for the Landsat data continuity mission. *IEEE Transactions on Geoscience and Remote Sensing* 51(2): 723–731. doi:10.1109/TGRS.2012.2203823.

Woodcock, C. E. et al. 2008. Free access to Landsat imagery, *Science* 320: 1011–1012. doi:10.1126/science.320.5879.1011a.

Wulder, M. A., Masek, J. G., Cohen, W. B., Loveland, T. R., and Woodcock, C. E. 2012. Opening the archive: How free data has enabled the science and monitoring promise of Landsat. *Remote Sensing of Environment* 122: 2–10. doi:10.1016/j.rse.2012.01.010.

Wulder, M. A., White, J. C., Loveland, T. R., Woodcock, C. E., Belward, A. S., Cohen, W. B., Fosnight, E. A., Shaw, J., Masek, J. G., and Roy, D. P. 2016. The global Landsat archive: Status, consolidation, and direction. *Remote Sensing of Environment* 185: 271–283. doi:10.1016/j.rse.2015.11.032.

Xian, G., Homer, C., and Fry, J. 2009. Updating the 2001 National Land Cover Database land cover classification to 2006 by using Landsat imagery change detection methods. *Remote Sensing of Environment* 113(6): 1133–1147. doi:10.1016/j. rse.2009.02.004.

Yuan, F., Sawaya, K. E., Loeffelholz, B. C., and Bauer, M. E. 2005. Land cover classification and change analysis of the Twin Cities (Minnesota) Metropolitan Area by multitemporal Landsat remote sensing. *Remote Sensing of Environment* 98(2): 317–328. doi:10.1016/j.rse.2005.08.006.

Zhang, Y., Guindon, B., and Cihlar, J. 2002. An image transform to characterize and compensate for spatial variations in thin cloud contamination of Landsat images. *Remote Sensing of Environment* 82(2): 173–187. doi:10.1016/S0034-4257(02)00034-2.

Zheng, D., Rademacher, J., Chen, J., Crow, T., Bresee, M., Le Moine, J., and Ryu, S. R. 2004. Estimating aboveground biomass using Landsat 7 ETM+ data across a managed landscape in northern Wisconsin, USA. *Remote Sensing of Environment* 93(3): 402–411. doi:10.1016/j.rse.2004.08.008.

Zhou, G., Zhou, X., Yue, T., and Liu, Y. 2016. An optional threshold with SVM cloud detection algorithm and DSP implementation. *ISPRS-International Archives of the Photogrammetry, Remote Sensing and Spatial Information Sciences* 41: 771–777. doi:10.5194/isprs-archives-XLI-B8-771-2016.

Zhu, Z. and Woodcock, C. E. 2012. Object-based cloud and cloud shadow detection in Landsat imagery. *Remote Sensing of Environment* 118: 83–94. doi:10.1016/j. rse.2011.10.028.

Zhu, Z. and Woodcock, C. E. 2014. Automated cloud, cloud shadow, and snow detection in multitemporal Landsat data: An algorithm designed specifically for monitoring land cover change. *Remote Sensing of Environment* 152: 217–234. doi:10.1016/j.rse.2014.06.012.

Zhu, Z., Wang, S., and Woodcock, C. E. 2015. Improvement and expansion of the Fmask algorithm: Cloud, cloud shadow, and snow detection for Landsats 4–7, 8, and Sentinel 2 images. *Remote Sensing of Environment* 159: 269–277. doi:10.1016/j. rse.2014.12.014.

2

An Automatic System for Reconstructing High-Quality Seasonal Landsat Time Series

Xiaolin Zhu, Eileen H. Helmer, Jin Chen, and Desheng Liu

CONTENTS

2.1 Introduction 25
2.2 Methods 28
    2.2.1 Classify Uncontaminated Pixels in Each Image.........................29
    2.2.2 Select Ancillary Data for Each Contaminated Pixel from
        the Time Series 29
    2.2.3 Interpolate Contaminated Pixels by NSPI................................30
2.3 Experiments 33
2.4 Results 35
    2.4.1 Reconstruction of Real Landsat Time Series 35
    2.4.2 Reconstruction of Simulated Landsat Time Series 36
2.5 Conclusion and Discussions 38
Acknowledgments 40
References 41

2.1 Introduction

Seasonal time series data from satellites are highly desired by researchers from different fields to study our Earth system. Seasonal time series data contain the temporal aspects of natural phenomena on the land surface, which are extremely helpful for discriminating different land cover types (Zhu and Liu, 2014), monitoring vegetation dynamics (Shen et al., 2011), estimating crop yields (Johnson et al., 2016), assessing environmental threats (Garrity et al., 2013), exploring human-nature interactions (Zhu and Woodcock, 2014a), and revealing ecology-climate feedbacks (Piao et al., 2015).

Since 2008, all Landsat images, archived and newly acquired, have been available at no charge for end users, stimulating the studies of land surface dynamics with seasonal Landsat time series, because these data have a spatial resolution appropriate for heterogeneous land surfaces, such as urban areas (Schneider, 2012; Zhou et al., 2014). However, like all optical satellite images, Landsat images are contaminated by clouds and cloud shadows (Ju and Roy,

2008). In addition, since May 2003, the scan-line corrector (SLC) of the Enhanced Thematic Mapper plus sensor (ETM+) on Landsat 7 has failed permanently. It causes roughly 22% of the pixels to be unscanned in any ETM+ image (referred to as SLC-off images) after 2003 (Chen et al., 2011). These sources of image contamination (i.e., clouds, cloud shadows, and SLC gaps) severely hinder Landsat time series applications and thus must be removed and then filled with data predicted from a different date, from nearby pixels or from both to reconstruct a high quality seasonal time series. Image reconstruction is especially important in cloudy regions such as tropical, subtropical, and high altitude regions.

Generally, reconstructing high quality Landsat time series involves two steps: screening clouds and cloud shadows and interpolating contaminated pixels. For screening clouds and cloud shadows in Landsat images, some promising methods have been developed, including Function of mask (Fmask) (Zhu and Woodcock, 2012), time series filtering method (Goodwin et al., 2013), and multiTemporal mask (Tmask) (Zhu and Woodcock, 2014b). Now, Fmask is used by the United States Geological Survey (USGS) to produce a cloud mask layer for end users. This product can be directly used when reconstructing Landsat time series. For interpolating contaminated pixels in Landsat images, existing techniques can be grouped into three categories: spatial interpolators (Cheng et al., 2014; Meng et al., 2009), temporal interpolators (Melgani, 2006; Zhu et al., 2015), and spatiotemporal interpolators (Chen et al., 2011; Zhu et al., 2012a,b).

Spatial interpolators search for one or several uncontaminated pixels that likely have the same land cover and that have very similar spectral values as the contaminated pixel, referred to as similar pixels. This searching process requires cloud-free ancillary images acquired at different times. Then, the contaminated pixel is replaced by similar pixels. Meng et al. (2009) developed a closest spectral fit (CSF) method to replace spectral values of cloudy pixels by cloud-free pixels using location-based one-to-one correspondence and spectral-based closest fit. Cheng et al. (2014) applied a pixel-offset based Markov random field global function to find the most suitable similar pixels to replace contaminated pixels.

Temporal interpolators estimate values of contaminated pixels through the temporal relationship among images acquired at different time points. In general, there are two ways to model the temporal relationship: as a function of cloud-free observations in the time series sequence (Melgani, 2006) or as a function of observation time (Zhu et al., 2015). To interpolate cloudy pixels in one image of a time series, the contextual multiple linear prediction (CMLP) method developed by Melgani (2006) uses a multiple linear function to model the relationship between the pixel values of the image being reconstructed and other images in the sequence. CMLP can only interpolate cloudy images one at a time in the sequence because it needs to build the temporal model for each cloudy image. Zhu et al. (2015) employed harmonic models to fit the temporal pattern of each pixel in a time series sequence. The temporal model is a function of the satellite image acquisition time, and its coefficients are estimated from all cloud-free points in the time series sequence. This trained temporal model delineates the seasonality of each pixel. It can be used to interpolate cloudy pixels in all images of the time series simultaneously.

Spatiotemporal interpolators integrate both spatial and temporal information into the prediction of contaminated pixels (Chen et al., 2011; Zhu et al., 2012a,b). Chen et al. (2011) proposed a Neighborhood Similar Pixel Interpolator (NSPI) method to interpolate the gaps caused by the SLC problem. NSPI first searches for similar pixels and then it uses these similar pixels as samples to predict gap pixels in two ways. A spatial prediction is obtained from a weighted average of similar pixel values in the target image, that is, the image to be constructed, to predict the gap pixels. A temporal prediction is obtained by retrieving the temporal change from similar pixel values in both the target image and ancillary images. Then the two predictions are combined based on confidence measures to get the final prediction of gap pixels. The NSPI method was further modified (i.e., MNSPI) for restoring the spectral values of cloudy pixels by considering the difference between narrow wedge-shaped SLC-off gaps and cloud, that is, clouds are randomly shaped clusters and most are much larger than SLC-off gaps (Zhu et al., 2012a). The temporally predicted result is given an increasingly larger weight as the spatial distance of the pixel from the cloud edge increases. The NSPI method was further improved through incorporating Geostatistics, which can help to estimate the uncertainties of interpolation results and reduce the number of empirical parameters (Zhu et al., 2012b).

In general, spatiotemporal interpolators have more advantages compared with either spatial interpolators or temporal interpolators alone. Limitations of spatial interpolators include: (1) accuracy decreases fast with the size of cloud patches; (2) spatial continuity of interpolated images cannot be well reserved in heterogeneous landscapes. Spatiotemporal interpolators use the corresponding pixel information from cloud-free ancillary images, so it is less affected by the size of cloud patches. The weighted combination of two predictions in spatiotemporal interpolators ensures the spatial and radiometric continuity of the reconstructed image in heterogeneous landscapes (Chen et al., 2011). Limitations of temporal interpolators are high uncertainties when applied in very cloudy regions. Temporal interpolators use cloud-free points in the time series to model a temporal pattern for each pixel, then this temporal pattern is used to predict values of cloudy points. However, in areas with persistent clouds, such as tropical regions, there are no adequate cloud-free points in the time series to build a reliable temporal model. Thus, it leads to large errors in the reconstructed seasonal time series. In contrast, spatiotemporal interpolators also use spatial information which is a valuable complement when the temporal information is not reliable.

However, the original NSPI and MNSPI were developed for interpolating gap pixels and cloudy pixels, respectively, in individual images. They are not efficient for reconstructing seasonal Landsat time series when clouds, cloud shadows, and SLC-off gaps exist in most images in a time series. To this end, this chapter introduces an automatic system for interpolating all types of contaminated pixels in all Landsat images of a time series through integrating NSPI and MNSPI into an iterative process. The input data of this system are a time series of Landsat images, along with associated cloud and cloud shadow masks, and the output is a time series with the same Landsat images but without missing pixels caused by clouds, cloud shadows, and SLC-off gaps. This system will promote the use of Landsat time series to monitor land surface dynamics.

2.2 Methods

Figure 2.1 shows the flowchart of the proposed automatic system for reconstructing Landsat time series. It requires two input data sets. First, multi-temporal original Landsat images are collected to composite a time series. These original Landsat images can be either the Level 1 Terrain corrected (L1T) product or land surface reflectance product downloaded from the USGS Earth Explorer system. If a Landsat image is covered by a lot of clouds, cloud shadows, and SLC-off gaps, its useful information is very limited. Therefore, only Landsat images with more than 40% clear pixels of the entire scene are included in the time series. All these selected Landsat images are stacked by their acquisition dates. Considering that original Landsat images have

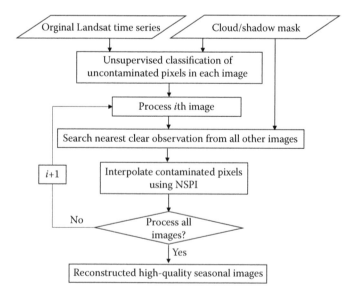

FIGURE 2.1
Flowchart of the automatic system for reconstructing high-quality seasonal Landsat time series.

been geo-rectified and calibrated accurately (Zhu and Woodcock, 2014b), no further preprocessing steps are needed. Second, a mask product is needed to indicate the contaminated pixels in the time series. Cloud masks of Landsat images are available from the USGS Earth Explorer system. Although the cloud mask produced by the Fmask algorithm has acceptable accuracy in many places, the omission of clouds and cloud shadows will lead to errors in the interpolation results. Therefore, two extra steps are used to reduce the omission errors in cloud mask: (1) considering that the omission errors often happen at the edges of cloud patches, a buffer of 1 or 2 pixels can be added to the cloud patches; (2) manual editing is needed if there are still contaminated pixels not marked in the mask. As Figure 2.1 shows, there are 3 main steps for the proposed automatic system. The details will be described below.

2.2.1 Classify Uncontaminated Pixels in Each Image

The NSPI method assumes that pixels with the same class should have high spectral similarity and similar temporal changing pattern (Chen et al., 2011). To speed up the searching of spectrally similar pixels in the original NSPI method, all input images in the time series are classified based on the spectral similarity of pixels. Here, to ensure the automation of the whole system, we use a classic unsupervised classifier, K-means, to classify the uncontaminated pixels in each image. The k-means method uses an iterative procedure. At each iteration, each pixel is assigned to one class based on the closeness with the class means obtained from the last iteration, and then class means are updated using a new class labels of pixels. The iterative process will be ended when the class labels no longer change (Lloyd, 1982). Number of classes, k, is an important parameter in K-means classification. It can be determined empirically based on the study area and the same value is used for all of the images. In most cases, the number of classes ranges from 3 to 10. Specifically, urban areas often have 7–10 land cover types, rural areas have 4–6 land cover types, and areas without human activities have 3–4 land cover types. The final result is a classification map of uncontaminated pixels for each of the original Landsat images.

2.2.2 Select Ancillary Data for Each Contaminated Pixel from the Time Series

In the NSPI method, to interpolate the value of a contaminated pixel, which is named as a target pixel, a clear observation in other images corresponding to this target pixel is needed. This clear observation is named as ancillary data. As a result, to interpolate all contaminated pixels in any one image of the time series, we need to select the best ancillary data from the time series for each contaminated pixel. Assuming that there are K images in the time series, we are interpolating the ith image ($i = 1, \ldots, K$). The searching process of ancillary data for all contaminated pixels in the ith image is as follow: (1) sort all other K-1 images in the time series based on the temporal closeness to the ith image; (2) start from the image closest to the ith image, clear pixels in this image corresponding to contaminated pixels in the ith image are selected as ancillary data of these contaminated pixels; (3) for the remaining contaminated pixels without ancillary data, do the above selecting process in the next closest image; (4) repeat the searching process until all contaminated pixels have ancillary data selected.

2.2.3 Interpolate Contaminated Pixels by NSPI

There are some differences between the NSPI method for SLC-off gap filling (Chen et al., 2011) and MNSPI for cloud removal (Zhu et al., 2012a). To integrate NSPI and MNSPI in one system for interpolating pixels contaminated by both clouds and SLC-off gaps, the original NSPI method needs some modifications. Below will briefly describe the steps of NSPI interpolation, while more details can be find in Chen et al. (2011) and Zhu et al. (2012a).

Let us assume that a contaminated pixel (x, y) in a Landsat image to be interpolated was acquired at t_2, and its corresponding ancillary data were provided by a Landsat image acquired at t_1. The first step is searching for similar pixels based on the ancillary data. The original NSPI and MNSPI use different strategies to search for similar pixels based on the different properties of SLC-off gaps and clouds. Here, a new strategy appropriate for both gaps and clouds is used. As shown in Figure 2.2, a target pixel contaminated by clouds is to be interpolated. First, using the spectral classification map of

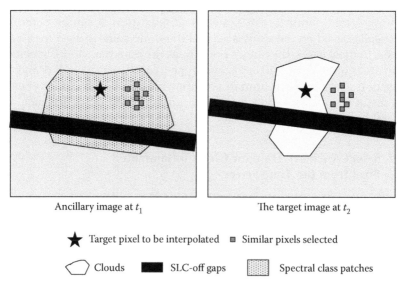

Ancillary image at t_1 The target image at t_2

★ Target pixel to be interpolated   ▪ Similar pixels selected

◯ Clouds   ▬ SLC-off gaps   ▦ Spectral class patches

FIGURE 2.2
Schematic diagram of the similar pixel selection.

the ancillary image, pixels that are clear in both the target image and the ancillary image and that are the same class as the target pixel are selected by an adaptive window procedure (Chen et al., 2011) as the candidates for similar pixels. Second, the N (e.g., 20) pixels that are most similar to the target pixel are selected as similar pixels from the candidates. Here, the similarity is defined as the root mean square deviation ($RMSD$) between each candidate pixel and the target pixel as Equation 2.1:

$$RMSD_i = \sqrt{\frac{\sum_{b=1}^{n}(L(x_i,y_i,t_1,b)-L(x,y,t_1,b))^2}{n}}, \qquad (2.1)$$

where $L(x_i, y_i, t_1, b)$ is the band b value of ith candidate pixel located in (x_i, y_i) in the ancillary image acquired at t_1, $L(x, y, t_1, b)$ and has the same definition but for a target pixel, and n is the number of spectral bands.

The similar pixels are given different weights when they are used to predict the value of the target pixel. The weight W_j determines the contribution of the jth similar pixel for predicting the value of the target pixel. This is determined by the spatial distance and the spectral similarity between the similar pixel and the target pixel (Gao et al., 2006). Higher spectral similarity and smaller distance of a similar pixel to the target pixel will increase the weight of that given pixel. Therefore, the weight of the jth similar pixel, W_j, can be calculated by the following equations:

$$D_j = \sqrt{(x_j - x)^2 + (y_j - y)^2}, \qquad (2.2)$$

$$CD_j = RMSD_j \times D_j, \qquad (2.3)$$

$$W_j = \frac{(1/CD_j)}{\sum_{j=1}^{N}(1/CD_j)}. \qquad (2.4)$$

The range of W_j is from 0 to 1, and the sum of all similar pixel weights is 1.

Then, two initial predictions using the spatial information and temporal information can be calculated, respectively. First, since the similar pixels have the same or approximate spectral value as the target pixel when they are observed at the same time, we can use the information of these similar pixels in the image at t_2 to predict the target pixel. Accordingly, the weighted average of all the similar pixels in the target image is used to make the first prediction for the target pixel:

$$L_1(x,y,t_2,b) = \sum_{j=1}^{N} W_j \times L(x_j,y_j,t_2,b). \qquad (2.5)$$

Secondly, for the target pixel, the value at t_2 equals the sum of the value at t_1 and the change from t_1 to t_2. Because the value at t_1 can be obtained directly from the ancillary image, we only need to estimate the change of the target pixel from t_1 to t_2. It is reasonable to assume that the change of similar pixels can represent the change of the target pixel, because the similar pixels have the same temporal pattern as the target pixel. Accordingly, the weighted average of the change provided by all the similar pixels is used to calculate the value of the target pixel as the second prediction:

$$L_2(x,y,t_2,b) = L(x,y,t_1,b) + \sum_{j=1}^{N} W_j \times (L(x_j,y_j,t_2,b) - L(x_j,y_j,t_1,b)). \quad (2.6)$$

Last, a weighted combination of the two initial predictions is used to compute the final prediction. The weights (T_1 and T_2) are determined by the extent of spatial continuity and the extent of temporal continuity between the ancillary image and the target image estimated from similar pixels. Here, the averaged $RMSD$ (R_1) between the similar pixel and the target pixel is used to denote the extent of the spatial continuity:

$$R_1 = \frac{1}{N} \sum_{j=1}^{N} \sqrt{\frac{\left[\sum_{b=1}^{n}(L(x_j,y_j,t_1,b) - L(x,y,t_1,b))^2\right]}{n}}. \quad (2.7)$$

In the same way, the averaged $RMSD$ (R_2) of similar pixels between observations at t_1 and t_2 is used to denote the extent of temporal continuity between the input image and the target image:

$$R_2 = \frac{1}{N} \sum_{j=1}^{N} \sqrt{\frac{\left[\sum_{b=1}^{n}(L(x_j,y_j,t_1,b) - L(x_j,y_j,t_2,b))^2\right]}{n}}. \quad (2.8)$$

Then, the normalized reciprocal of R_1 and R_2 is used as the weight T_1 and T_2 respectively:

$$T_i = \frac{(1/R_i)}{(1/R_1 + 1/R_2)}, \quad \text{where} \quad i = 1,\ 2. \quad (2.9)$$

The final predicted value of the target pixel is calculated as:

$$L(x,y,t_2,b) = T_1 \times L_1(x,y,t_2,b) + T_2 \times L_2(x,y,t_2,b) \quad (2.10)$$

The above interpolation process is implemented for all contaminated pixels in each image of the time series until all images are reconstructed.

2.3 Experiments

To evaluate the feasibility and effectiveness of the proposed automatic system, two experiments were carried out: a real cloudy time series reconstruction and a simulated cloudy time series reconstruction.

For the experiment with the real cloudy time series reconstruction, 15 Landsat-7 images covering Mona Island of Puerto Rico (a size of 400×400 pixels) were used. Mona Island lies 66 km west of Puerto Rico. It is a mainly flat limestone plateau surrounded by sea cliffs and covered by tropical dry forests. As is typical in tropical regions, most Landsat images of this island are cloudy. In this experiment, we used all 15 Landsat-7 images acquired from 2008 and 2009 with less than 60% of pixels contaminated by clouds, cloud shadows, and SLC-off gaps. These images were stacked according to their dates (i.e., day of year) to composite a one-year time series (Table 2.1). From Table 2.1, the proportion of contaminated pixels in images varies from 23.67% to 57.29%. Figure 2.3 shows a false-color composite of 3 images with low, medium, and high proportions of contaminated pixels. It is a common issue for existing methods that the accuracy of image reconstruction decreases with

TABLE 2.1

Summary of Landsat-7 Images in Mona Island Site

ID	Year	Day of Year	Contaminated Pixels %
1	2008	11	26.37
2	2008	27	25.96
3	2008	43	25.25
4	2009	61	32.89
5	2009	77	37.62
6	2008	91	25.68
7	2009	109	46.68
8	2008	139	57.29
9	2008	171	48.47
10	2008	203	27.48
11	2009	253	53.11
12	2009	301	24.23
13	2009	317	46.68
14	2009	333	27.63
15	2009	349	23.67

FIGURE 2.3
(**See color insert.**) False-color composite of 3 Landsat-7 images in Mona Island with low, medium, and high proportion of contaminated pixels (from left to right).

the proportion of missing pixels, so the different proportions of contaminated pixels in this site are good for testing whether or not the performance of the proposed method is acceptable for images with large proportions of contaminated pixels.

To more quantitatively assess the accuracy of seasonal image reconstruction, another study site with a lot of cloud-free Landsat images was used to implement a simulation study. Fifteen cloud-free Landsat 7 images were provided by Emelyanova et al. (2013). This site is in southern New South Wales, Australia, and has a heterogeneous landscape. The major land cover types in this area are irrigated rice cropland, dryland agriculture, and woodlands. These images were acquired during 2001 October to 2002 May. Rice croplands are often irrigated in October-November, which leads to large temporal changes in the time series. Both high temporal change and high heterogeneity challenge the reconstruction of contaminated pixels, so this data set is ideal for testing the effectiveness of the proposed method. We cut all images to the image size of Mona Island for this experiment. Cloud masks from the Mona Island site were overlaid onto all 15 cloud-free images in this site to produce pseudo cloudy images. Figure 2.4 shows one image in this site and its corresponding simulated contaminated image using the mask from a Mona Island image shown in Figure 2.3 (the right one).

Then, the proposed method was applied to these pseudo cloudy images. The accuracy was evaluated through the calculation of two statistical indices. The first index is the root mean square error ($RMSE$). This metric is frequently used to assess the differences between values predicted by a model and the values observed or measured. A larger $RMSE$ indicates a larger prediction error. The second metric is correlation coefficient (R) between the actual values and predicted values of contaminated pixels. R is used to show the linear relationship between actual and reconstructed images. An R value closer to 1 indicates higher accuracy of reconstructed images.

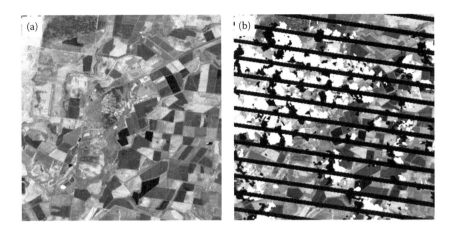

FIGURE 2.4
A Landsat-7 image acquired on February 11, 2002 (a) and its simulated contaminated image using the mask from the Mona Island image of 2008 DOY139 (b).

2.4 Results

2.4.1 Reconstruction of Real Landsat Time Series

All the 15 images in the time series were reconstructed successfully. To save space here, Figure 2.5 only shows the reconstructed results of three images in Figure 2.3, but other images have similar reconstruction results. We can see that regardless of the proportion of contaminated pixels the images have, the reconstructed images appear spatially continuous without displaying the footprints of gaps or cloud/shadow patches. The result suggests that the proposed automatic system can interpolate contaminated pixels accurately to produce high-quality time series.

Although we cannot assess the accuracy quantitatively for Mona, because the true pixel values are unknown, we can analyze the time series based

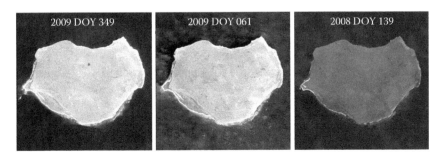

FIGURE 2.5
(See color insert.) Reconstruction results of three contaminated images shown in Figure 2.3.

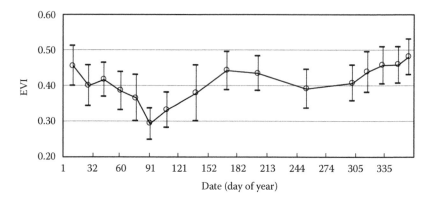

FIGURE 2.6
Enhanced vegetation index (EVI) time series of Mona Island derived from reconstructed 15 Landsat-7 images. Each circle is the mean value of all pixels in this study area and the error bar is one standard deviation.

on the prior knowledge of the vegetation growing cycle to evaluate the usefulness of the reconstructed images. Figure 2.6 shows the average enhanced vegetation index (EVI) time series (Huete et al., 2002) derived from the 15 reconstructed images over Mona Island. The dry season of Mona Island is from January to April. As can be seen from the EVI curve, the EVI values decrease during the dry season and reach the lowest point in the end of March. This temporal pattern is consistent with the local vegetation pattern as the deciduous tree species lose leaves during the dry season. The EVI curve from the reconstructed time series indicates the reliability of the proposed method. It can help us to monitor vegetation seasonality at high frequency.

2.4.2 Reconstruction of Simulated Landsat Time Series

Quantitative accuracy assessment of reconstruction results of all 15 simulated contaminated images are reported in Table 2.2. The proposed method was used to reconstruct all 6 bands with 30-m resolution. To save space and because results from the two shortwave bands are similar to those for the first four bands, only results from the first 4 bands were listed in Table 2.2. R values of all images are higher than 0.8, and most RMSE values are less than 0.02, suggesting that the proposed automatic system interpolated all contaminated pixels very accurately. In addition, most existing methods do not usually yield acceptable results when there are large proportions of contaminated pixels. Figure 2.7 shows the scatter plot between the accuracy (both RMSE and R) of near infrared (NIR) band and the proportion of contaminated pixels. No clear relationship exists in the scatter plot, suggesting that the proposed method yields a robust result, regardless the proportion of pixels contaminated in an image.

TABLE 2.2

RMSE and R Values of Reconstructed 15 Simulated Landsat-7 Cloudy Images

				RMSE				R			
ID	Year	DOY[a]	CP %[b]	Band1	Band2	Band3	Band4	Band1	Band2	Band3	Band4
1	2001	17	26.37	0.0075	0.0092	0.0141	0.0300	0.90	0.90	0.91	0.96
2	2001	33	25.96	0.0080	0.0104	0.0152	0.0304	0.88	0.88	0.90	0.95
3	2001	40	25.25	0.0081	0.0105	0.0165	0.0359	0.84	0.84	0.86	0.90
4	2001	56	32.89	0.0073	0.0096	0.0145	0.0230	0.91	0.92	0.93	0.94
5	2001	65	37.62	0.0098	0.0146	0.0231	0.0319	0.89	0.90	0.91	0.91
6	2002	97	25.68	0.0081	0.0117	0.0183	0.0273	0.96	0.96	0.97	0.93
7	2002	104	46.68	0.0094	0.0142	0.0219	0.0285	0.94	0.94	0.95	0.92
8	2002	136	57.29	0.0098	0.0131	0.0193	0.0281	0.91	0.91	0.92	0.91
9	2002	145	48.47	0.0092	0.0118	0.0182	0.0296	0.89	0.89	0.90	0.91
10	2002	168	27.48	0.0082	0.0109	0.0148	0.0250	0.90	0.88	0.92	0.95
11	2002	184	53.11	0.0074	0.0089	0.0130	0.0255	0.91	0.90	0.93	0.94
12	2002	193	24.23	0.0082	0.0108	0.0162	0.0269	0.85	0.87	0.89	0.92
13	2002	200	46.68	0.0088	0.0112	0.0158	0.0262	0.84	0.85	0.88	0.91
14	2002	209	27.63	0.0075	0.0086	0.0118	0.0164	0.91	0.93	0.94	0.97
15	2002	216	23.67	0.0065	0.0078	0.0114	0.0177	0.91	0.94	0.95	0.96

[a] DOY: days starting from October 1, 2001.
[b] CP%: proportion of simulated contaminated pixels.

Figure 2.8 shows the NIR band results of two simulated contaminated images. One image has a low proportion of contaminated pixels (upper row in Figure 2.8), while another one has a high proportion of contaminated pixels (lower row in Figure 2.8). By visual comparison of both of the reconstructed images shown in Figure 2.8, the reconstructed images are very close to the actual images. We can see that all the spatial details were retrieved well,

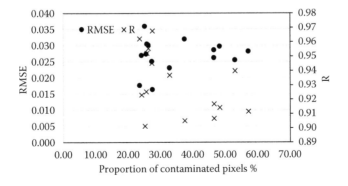

FIGURE 2.7
Scatter plot between proportion of contaminated pixels in each image and the accuracy of reconstructed images: root mean square error (RMSE) and correlation coefficient R.

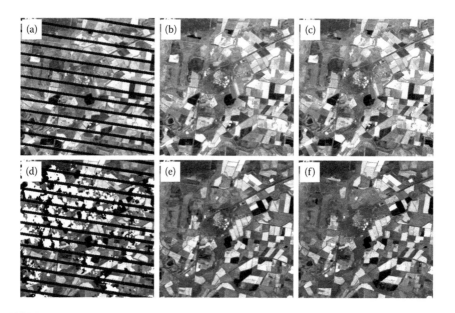

FIGURE 2.8
Images in upper row are NIR band of simulated contaminated Landsat-7 image of DOY*136 (a), the true image (b), and the reconstructed image (c). Images in lower row are NIR band of simulated contaminated Landsat-7 image of DOY*216 (d), the true image (e), and the reconstructed image (f).

indicating that the contaminated pixels interpolated by our method have high accuracy, even when the landscape is very heterogeneous.

Figure 2.9 shows the average NDVI time series derived from the 15 cloud-free Landsat-7 images compared with the time series as reconstructed with the simulated contaminated images. It is clear that the two NDVI temporal curves are very consistent with each other. Both curves describe well the temporal dynamics associated with crop phenology over a single growing season. During the irrigation period, the NDVI values decrease. Then, NDVI values increase during the crop growing period (Emelyanova et al., 2013). This temporal pattern from reconstructed images shows that the reconstructed time series can be used to monitor a fast-changing agricultural landscape.

2.5 Conclusion and Discussions

Landsat time series data have been widely used to study seasonal land surface dynamics at regional and global scales, but the SLC failure in Landsat-7 and cloud contamination reduce the chances of obtaining a

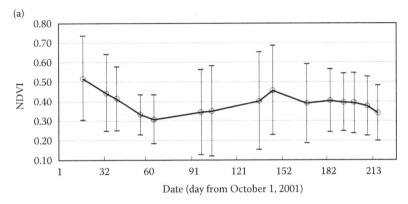

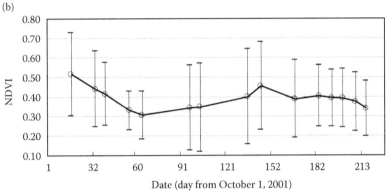

FIGURE 2.9
NDVI time series derived from original cloud-free Landsat-7 images (a) and from reconstructed
results of simulated contaminated images (b). Each circle is the mean value of all pixels and the
error bar is one standard deviation.

high-quality time series. Therefore, the necessary and feasible way to
address this problem is to interpolate contaminated pixels to reconstruct the
time series. This chapter introduces an automatic system to reconstruct all
contaminated images simultaneously in a time series, which is based on the
NSPI method and an iterative process. The real and simulation experiments
show that the proposed method can reconstruct dense Landsat time series
successfully. The vegetation indices derived from the reconstructed time
series can be used to monitor the vegetation seasonality, including forest
phenology and crop growing stages. The robustness of the proposed method
for interpolating images with large proportions of missing pixels is very
important because it is difficult to acquire good quality images in cloudy
regions (Ju and Roy, 2008). Considering that the Landsat series have collected
data over 40 years, the method introduced in this chapter will promote
the use of seasonal Landsat time series in various disciplines, including
ecology, hydrology, earth science, environmental science, agriculture, and

even sociology. Studying seasonality can advance our understanding of interactions between different Earth systems and improve our ability to predict future scenarios.

In general, the proposed method has the following strengths: (1) it makes full use of temporal and spatial information contained in all images in the time series and produces the contamination-free images in all seasons simultaneously; (2) it integrates cutting edge techniques for SCL-off gap filling and cloud removal into one system so that various types of missing pixels can be processed at the same time; (3) it is highly automated with minimal predefined parameters and human-computer interactions, so it can be user friendly for people from fields other than remote sensing. The proposed method has been successfully applied to study vegetation phenology in Puerto Rico and the US Virgin Islands (Gwenzi et al., 2017). The open-source program of the proposed method can be downloaded from the developer's website: xiaolinzhu.weebly.com.

The proposed method may face some challenges in some specific situations. First, in regions with extremely persistent clouds, it would be very difficult to reconstruct Landsat time series with acceptable accuracy because clouds totally cover many Landsat images. A possible way to solve this problem is to incorporate data from other sensors with short revisit cycles, such as MODerate resolution Imaging Spectroradiometer (MODIS) images. Currently, some spatiotemporal data fusion methods have been developed to fuse MODIS and Landsat images to increase the frequency of Landsat observations (Gao et al., 2006; Zhu et al., 2016, 2010). These techniques could be integrated into our automatic system in the future. Second, the proposed method contains a similar pixel searching process for each contaminated pixel, so it may need a lot of computing time when processing massive images over large area. Therefore, we recommend using high performance computers or parallel computing to increase the computing speed when it is used to reconstruct Landsat time series over large area and long period.

Acknowledgments

This study was supported by the Research Grants Council of Hong Kong (project no. 25222717), the National Natural Science Foundation of China (project no. 41701378), the USDA Forest Service International Institute of Tropical Forestry (Cooperative Agreement 13-CA-11120101-029), the Southern Research Station Forest Inventory and Analysis Program, the National Science Foundation of the USA (1010314), and it was conducted in cooperation with the University of Puerto Rico and the Rocky Mountain Research Station.

References

Chen, J., Zhu, X., Vogelmann, J.E., Gao, F., Jin, S., 2011. A simple and effective method for filling gaps in Landsat ETM+ SLC-off images. *Remote Sens. Environ.* 115, 1053–1064.

Cheng, Q., Shen, H., Zhang, L., Yuan, Q., Zeng, C., 2014. Cloud removal for remotely sensed images by similar pixel replacement guided with a spatio-temporal MRF model. *ISPRS J. Photogramm. Remote Sens.* 92, 54–68. doi:10.1016/j.isprsjprs.2014.02.015

Emelyanova, I. V., McVicar, T.R., Van Niel, T.G., Li, L.T., van Dijk, A.I.J.M., 2013. Assessing the accuracy of blending Landsat-MODIS surface reflectances in two landscapes with contrasting spatial and temporal dynamics: A framework for algorithm selection. *Remote Sens. Environ.* 133, 193–209. doi:10.1016/j.rse.2013.02.007

Gao, F., Masek, J., Schwaller, M., Hall, F., 2006. On the Blending of the Landsat and MODIS Surface Reflectance: Predicting Daily Landsat Surface Reflectance. *IEEE Trans. Geosci. Remote Sens.* 44, 2207–2218.

Garrity, S.R., Allen, C.D., Brumby, S.P., Gangodagamage, C., McDowell, N.G., Cai, D.M., 2013. Quantifying tree mortality in a mixed species woodland using multitemporal high spatial resolution satellite imagery. *Remote Sens. Environ.* 129, 54–65. doi:10.1016/j.rse.2012.10.029

Goodwin, N.R., Collett, L.J., Denham, R.J., Flood, N., Tindall, D., 2013. Cloud and cloud shadow screening across Queensland, Australia: An automated method for Landsat TM/ETM+ time series. *Remote Sens. Environ.* 134, 50–65. doi:10.1016/j.rse.2013.02.019

Gwenzi, D., Helmer, E.H., Zhu, X., Lefsky, M.A., Marcano-Vega, H., 2017. Predictions of tropical forest biomass and biomass growth based on stand height or canopy area are improved by Landsat-scale phenology across Puerto Rico and the US Virgin Islands. *Remote Sens.* 9(2), 1–18. doi:10.3390/rs9020123

Huete, A., Didan, K., Miura, T., Rodriguez, E.P., Gao, X., Ferreira, L.G., 2002. Overview of the radiometric and biophysical performance of the MODIS vegetation indices. *Remote Sens. Environ.* 83, 195–213. doi:10.1016/S0034-4257(02)00096-2

Johnson, M.D., Hsieh, W.W., Cannon, A.J., Davidson, A., Bédard, F., 2016. Crop yield forecasting on the Canadian Prairies by remotely sensed vegetation indices and machine learning methods. *Agric. For. Meteorol.* 218–219, 74–84. doi:10.1016/j.agrformet.2015.11.003

Ju, J., Roy, D.P., 2008. The availability of cloud-free Landsat ETM+ data over the conterminous United States and globally. *Remote Sens. Environ.* 112, 1196–1211. doi:10.1016/j.rse.2007.08.011

Lloyd, S.P., 1982. Least Squares Quantization in PCM. *IEEE Trans. Inf. Theory* 28, 129–137. doi:10.1109/TIT.1982.1056489

Melgani, F., 2006. Contextual reconstruction of cloud-contaminated multitemporal multispectral images. *IEEE Trans. Geosci. Remote Sens.* 44, 442–455. doi:10.1109/TGRS.2005.861929

Meng, Q., Borders, B.E., Cieszewski, C.J., Madden, M., 2009. Closest Spectral Fit for Removing Clouds and Cloud Shadows. *Photogramm. Eng. Remote Sens.* 75, 7505–7505. doi:10.14358/PERS.75.5.569

Piao, S., Yin, G., Tan, J., Cheng, L., Huang, M., Li, Y., Liu, R. et al., 2015. Detection and attribution of vegetation greening trend in China over the last 30 years. *Glob. Chang. Biol.* 21, 1601–1609. doi:10.1111/gcb.12795

Schneider, A., 2012. Monitoring land cover change in urban and peri-urban areas using dense time stacks of Landsat satellite data and a data mining approach. *Remote Sens. Environ.* 124, 689–704. doi:10.1016/j.rse.2012.06.006

Shen, M., Tang, Y., Chen, J., Zhu, X., Zheng, Y., 2011. Influences of temperature and precipitation before the growing season on spring phenology in grasslands of the central and eastern Qinghai-Tibetan Plateau. *Agric. For. Meteorol.* 151, 1711–1722.

Zhou, W., Qian, Y., Li, X., Li, W., Han, L., 2014. Relationships between land cover and the surface urban heat island: Seasonal variability and effects of spatial and thematic resolution of land cover data on predicting land surface temperatures. *Landsc. Ecol.* 29, 153–167. doi:10.1007/s10980-013-9950-5

Zhu, X., Chen, J., Gao, F., Chen, X., Masek, J.G., 2010. An enhanced spatial and temporal adaptive reflectance fusion model for complex heterogeneous regions. *Remote Sens. Environ.* 114, 2610–2623. doi:10.1016/j.rse.2010.05.032

Zhu, X., Gao, F., Liu, D., Chen, J., 2012a. A modified neighborhood similar pixel interpolator approach for removing thick clouds in landsat images. *IEEE Geosci. Remote Sens. Lett.* 9, 521–525. doi:10.1109/LGRS.2011.2173290

Zhu, X., Helmer, E.H., Gao, F., Liu, D., Chen, J., Lefsky, M.A., 2016. A flexible spatiotemporal method for fusing satellite images with different resolutions. *Remote Sens. Environ.* 172, 165–177. doi:10.1016/j.rse.2015.11.016

Zhu, X., Liu, D., 2014. Accurate mapping of forest types using dense seasonal landsat time-series. *ISPRS J. Photogramm. Remote Sens.* 96, 1–11.

Zhu, X., Liu, D., Chen, J., 2012b. A new geostatistical approach for filling gaps in Landsat ETM+ SLC-off images. *Remote Sens. Environ.* 124, 49–60.

Zhu, Z., Woodcock, C.E., 2012. Object-based cloud and cloud shadow detection in Landsat imagery. *Remote Sens. Environ.* 118, 83–94. doi:10.1016/j.rse.2011.10.028

Zhu, Z., Woodcock, C.E., 2014a. Continuous change detection and classification of land cover using all available Landsat data. *Remote Sens. Environ.* 144, 152–171. doi:10.1016/j.rse.2014.01.011

Zhu, Z., Woodcock, C.E., 2014b. Automated cloud, cloud shadow, and snow detection in multitemporal Landsat data: An algorithm designed specifically for monitoring land cover change. *Remote Sens. Environ.* 152, 217–234. doi:10.1016/j.rse.2014.06.012

Zhu, Z., Woodcock, C.E., Holden, C., Yang, Z., 2015. Generating synthetic Landsat images based on all available Landsat data: Predicting Landsat surface reflectance at any given time. *Remote Sens. Environ.* 162, 67–83. doi:10.1016/j.rse.2015.02.009

3

Spatiotemporal Data Fusion to Generate Synthetic High Spatial and Temporal Resolution Satellite Images

Jin Chen, Yuhan Rao, and Xiaolin Zhu

CONTENTS

3.1 Introduction 43
3.2 NDVI Linear Mixing Growth Model (NDVI-LMGM) 46
    3.2.1 Description of NDVI-LMGM 46
    3.2.2 Test Experiment 49
        3.2.2.1 Assessment over Spatial and Temporal
               Contrasting Regions 51
        3.2.2.2 Assessment for Long Term Prediction 53
3.3 Flexible Spatiotemporal Data Fusion Method (FSDAF) 54
    3.3.1 Description of FSDAF 54
    3.3.2 Test Experiment 58
        3.3.2.1 Assessment of Simulated Results 60
        3.3.2.2 Assessment of Fusing Real Satellite Images 60
3.4 Conclusions and Discussion 62
Acknowledgments 64
References 64

3.1 Introduction

Remote sensing data with high spatial resolution and frequent coverage are highly valuable for monitoring land surface dynamics, such as change detection (Chen et al., 2015; Lu et al., 2016), vegetation phenology monitoring (Shen et al., 2011), and drought monitoring (Gao et al., 2015). This type of data is needed for monitoring heterogeneous landscapes or rapid surface changes (Zhu et al., 2016). However, there is no single sensor able to provide such data due to trade-off between acquisition frequency and spatial resolution (Gao et al., 2006). Currently, time series data with frequent temporal coverage (usually 1–2 times per day) usually have relatively coarse spatial resolution (250–1000 m), such as data from MODerate resolution Imaging Spectrometer

(MODIS). These data cannot reflect spatial details in heterogeneous landscapes. On the contrary, available high spatial resolution data (10–30 m) usually have long revisit time (10–16 days). They can provide rich spatial details but fail to capture rapid changes and the timing of changes.

With fast growing needs for time series data with high spatial and temporal resolution, spatiotemporal data fusion methods have been developed to combine advantages of both types of data (see Figure 3.1). Existing spatiotemporal data fusion methods usually require one or more fine-coarse resolution data pairs for training and coarse resolution data at target dates as input data. The temporal change information from coarse time series and spatial information from fine-resolution data are combined through spatiotemporal data fusion methods. The synthetic data could support monitoring rapid surface dynamics over heterogeneous landscapes (Gao et al., 2015; Zhu et al., 2016).

Currently, existing spatiotemporal data fusion methods deal with different levels of remote sensing data. Most of the existing spatiotemporal data fusion methods are developed to fuse raw reflectance data (Gao et al., 2006; Zhu et al., 2010). The synthetic reflectance data could then be used for further applications, such as land cover classification and change detection. Other spatiotemporal data fusion methods are developed for fusing products derived directly from reflectance or radiance. One of the most popular products is Normalized Difference Vegetation Index (NDVI) (Busetto et al., 2008) since it is widely used for ecosystem monitoring, land cover classification, and other applications.

Although existing spatiotemporal fusion methods process different levels of remote sensing products (reflectance or other products), they can be divided into three groups: (1) weighted function based, (2) unmixing based, and (3) dictionary-pair learning based. Weighted function based methods

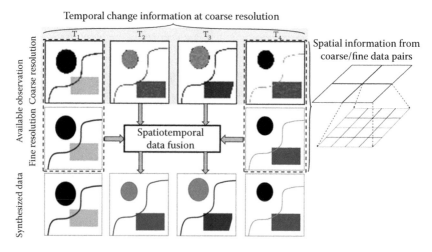

FIGURE 3.1
Schematic example of spatiotemporal data fusion to produce synthetic fine-resolution time series.

usually assume that temporal changes in "pure" coarse pixels are comparable with corresponding fine pixels. Therefore, changes of coarse pixels could be directly added to observed fine pixels for prediction. However, this cannot be satisfied for mixed pixels. Therefore, Gao et al. (2006) proposed the Spatial and Temporal Adaptive Reflectance Fusion Model (STARFM) which uses a weighted function assigning higher weights to purer coarse pixels based on information from neighboring fine pixels to generate synthetic data. Later, STARFM was modified for complex situations, resulting in the Spatial Temporal Adaptive Algorithm for mapping Reflectance CHange (STAARCH), which improves STARFM's performance when land cover type change and disturbances exist (Hilker et al., 2009), and another Enhanced STARFM (ESTARFM) (Zhu et al., 2010), which improves STARFM's accuracy in heterogeneous areas by introducing a conversion coefficient to adjust the change information from coarse images.

Unmixing based methods usually assume that the information of a coarse pixel is the weighted sum of information of different classes within the coarse pixel. Multisensor Multiresolution Techniques (MMT) (Zhukov et al., 1999) directly unmixes coarse pixels' reflectance using a least square method to produce synthetic reflectance time series. Recently, there have been several variations proposed to improve the accuracy of the unmixing based method. Zurita-Milla et al. (2008) use constraints during linear unmixing to ensure that synthetic data is within valid ranges. Amorós-López et al. (2013) focused on modifying the cost function of the unmixing process to avoid abnormal unmixed reflectance values. The Weighted Linear Mixture method (Busetto et al., 2008) predicts NDVI for fine pixels using a weighted linear unmixing method. However, fine resolution data are mostly used to provide fractional information of each endmember to unmix coarse resolution data at the prediction dates, which neglects abundant important spatial details in the fine resolution data.

Dictionary-pair learning based methods are relatively new. They establish correspondences between data of different resolutions based on their structural similarity. The Sparse-representation-based SpatioTemporal reflectance Fusion Model (SPSTFM) establishes a correspondence between the change of two pairs of fine- and coarse-resolution data through dictionary-pair learning, and then the trained dictionary is applied to predict fine data at the prediction dates (Huang and Song, 2012). Following SPSTFM, Song and Huang (2013) developed another dictionary-pair learning based fusion method which uses only one fine/coarse-resolution data pair for training, and then downscales the coarse image at the prediction date using a sparse coding technique.

The existing body of literature has demonstrated the capability of spatiotemporal data fusion methods to provide synthetic data with high spatial and temporal resolutions, which has been used for land cover classification (Chen et al., 2015), change detection (Lu et al., 2016), agriculture monitoring (Gao et al., 2015), and phenology detection (Shen et al., 2011). However, most of spatiotemporal data fusion methods face challenges caused by abrupt land cover type changes over heterogeneous landscape (Gao et al., 2015; Zhu et al.,

2016). Both weighted function based and unmixing based methods usually assume no land cover type change between prediction and observation dates. This assumption is usually invalid in heterogeneous landscapes, which leads to large prediction uncertainty under this condition. Most unmixing based methods also suffer from a block effect caused by within class variability and unrealistic predictions due to data noise and collinearity issues (Gevaert and García-Haro, 2015). For dictionary-pair learning based methods, the prediction is purely based on statistical relationship instead of physical properties of remote sensing data, which could distort the shape of objects in the fused images (Zhu et al., 2016).

In this chapter, we introduce two recently developed advanced spatio-temporal data fusion methods, the NDVI Linear Mixing Growth Model (NDVI-LMGM, (Rao et al., 2015)) and the Flexible Spatiotemporal DAta Fusion method (FSDAF, (Zhu et al., 2016)), both with improved accuracy. Although both methods are unmixing based methods, these two data fusion methods are developed for different purposes. NDVI-LMGM is specifically designed to construct high spatial and temporal resolution NDVI time series and to address local variability within the same land cover type. On the other hand, FSDAF is designed for reflectance data with the focus to improve the data fusion performance when abrupt land cover changes, such as floods and wildfires, occur during the data fusion period. For NDVI-LMGM, a complete coarse resolution time series with at least one high resolution data are required as input for data fusion, while FSDAF only requires at least one pair of fine/coarse resolution data and coarse resolution data at the prediction date. Due to the different purposes and implementation requirements of these two methods, this chapter will introduce both methods separately so readers can focus on methods based on different applications. The chapter will be organized as follow: the NDVI-LMGM will be first introduced in the Section 3.2, and FSDAF will be described in the Section 3.3. For each method, a comprehensive assessment will be presented, respectively. Finally, general comments on spatiotemporal data fusion methods and a comparison between FSDAF and NDVI-LMGM will be made in Section 3.4. This chapter can help readers to understand spatiotemporal data fusion methods for different features of interest, thus helping users to select an appropriate method for their specific applications based on the properties of NDVI-LMGM and FSDAF.

3.2 NDVI Linear Mixing Growth Model (NDVI-LMGM)

3.2.1 Description of NDVI-LMGM

Based on unmixing theory, NDVI-LMGM assumes that the temporal change of NDVI at coarse resolution is a weighted average of NDVI temporal changes of different classes (Equation 3.1),

$$\Delta NDVI^C(x,y) = \sum_{i=1}^{l} f_i(x,y) \times \Delta NDVI_c^F \tag{3.1}$$

where $\Delta NDVI^C(x, y)$ is the NDVI temporal change of a coarse pixel (x, y) for a time segment, $\Delta NDVI_i^F$ is the temporal change of NDVI for i-th class within the coarse pixel (x, y), and $f_i(x, y)$ is the fraction of the i-th class within (x, y).

With the fine resolution data or prior knowledge, the class fraction could be calculated based on classification using fine resolution data. Then, the temporal change of NDVI of each class at fine resolution could be estimated by solving the linear system described by Equation 3.2.

$$
\begin{bmatrix}
\Delta NDVI^C(x_1,y_1) \\
\vdots \\
\Delta NDVI^C(x,y) \\
\vdots \\
\Delta NDVI^C(x_n,y_n)
\end{bmatrix}
=
\begin{bmatrix}
f_1(x_1,y_1) & f_2(x_1,y_1) & \cdots & f_l(x_1,y_1) \\
\vdots & \vdots & & \vdots \\
f_1(x,y) & f_2(x,y) & \cdots & f_l(x,y) \\
\vdots & \vdots & & \vdots \\
f_1(x_n,y_n) & f_2(x_n,y_n) & \cdots & f_l(x_n,y_n)
\end{bmatrix}
\begin{bmatrix}
\Delta NDVI_1^F \\
\vdots \\
\Delta NDVI_i^F \\
\vdots \\
\Delta NDVI_l^F
\end{bmatrix}
\tag{3.2}
$$

To solve this linear system, at least l coarse resolution pixels should be used. It is reasonable to assume that the same classes within a local area share the same temporal change due to similar environmental conditions. Therefore, NDVI-LMGM uses a spatial moving window approach to solve fine resolution temporal changes for the center coarse pixel using a least square method (Rao et al., 2015). In practice, the size of the spatial moving window, which ranges from 3×3 to 5×5 coarse pixels, should be decided based on the complexity of the study area.

Figure 3.2 presents the general process of NDVI-LMGM. For NDVI-LMGM, it is necessary to make sure that input data are cloud free, otherwise cloud contamination will be passed to the synthetic output data. For coarse NDVI time series, such as MODIS data, its cloud contamination can be filtered using the Savitzky-Golay algorithm (Chen et al., 2004). For fine resolution data, such as Landsat data, cloud contamination is also inevitable for many regions (Ju and Roy, 2008). If fine-resolution data is partly covered by cloud, cloudy areas can be removed by the Neighborhood Similar Pixel Interpolator (NSPI) method proposed by Zhu et al. (2012) before NDVI-LMGM. Moreover, both fine and coarse images should be geo-registered using manually selected ground control points or some automated techniques, such as the automated registration and orthorectification package (AROP) (Gao et al., 2009).

After preprocessing, the fine-resolution data will be classified to estimate class fraction $f_i(x, y)$ for each coarse pixel. To keep the process automatic, Iterative Self-Organizing Data Analysis Technique (ISODATA), an unsupervised classification method, is used for classifying the fine-resolution

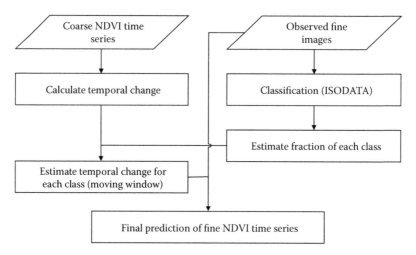

FIGURE 3.2
The flowchart of NDVI Linear Mixing Growth Model (NDVI-LMGM).

data. It should be noted that if ground reference data is available, supervised classification could be used to improve the classification accuracy. Moreover, if the high-resolution land cover map is available, it could be also used to replace the classification to calculate class fractions. Since multi-temporal fine-resolution data can improve classification accuracy (Guerschman et al., 2003; Rao et al., 2015), it is best to collect all available fine resolution data covering the study area for classification.

For a given time segment (from t_1 to t_2), when $f_i(x, y)$ has been determined by classification results, the temporal change in NDVI for each class within the target coarse pixel could be calculated by solving Equation 3.2 using a constrained least square method. The linear system described in Equation 3.2 uses information within a spatial moving window centered on the target pixel (Figure 3.3). NDVI-LMGM assumes that temporal change for each class is only similar within the local area, which could account for within class variability. To avoid unreasonable values caused by errors and noise when solving Equation 3.2, the constraints (Equation 3.3) are applied for all classes during unmixing,

$$\Delta NDVI_{min}^C - \sigma(\Delta NDVI^C) \leq \Delta NDVI_c^F \leq \Delta NDVI_{max}^C + \sigma(\Delta NDVI^C) \quad (3.3)$$

$\sigma(\Delta NDVI^C)$, $\Delta NDVI_{min}{}^C$, and $\Delta NDVI_{max}{}^C$ are the standard deviation, minimum, and maximum temporal change of all coarse pixels for the entire region.

After obtaining temporal changes for each class within coarse pixels, the prediction at t_2 could be made by adding temporal changes of each class to observed values at t_1. Moreover, if there are two available fine-resolution data from different observation dates (i.e., $t_{1,m}$ and $t_{1,n}$), a weighted combination

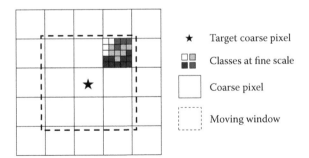

FIGURE 3.3
Schematic figure of composing a linear system using a spatial moving window.

procedure will be implemented to integrate predictions using two prediction results P_m and P_n to produce the final prediction,

$$P w_m P_m w_n P_n = + ,$$

where

$$w_k = \frac{1 / \left| \sum\limits_{i=-s}^{s} NDVI^C(x+i,y+i,t_{1,k}) - \sum\limits_{i=-s}^{s} NDVI^C(x+i,y+i,t_2) \right|}{\sum\limits_{k=m,n} 1 / \left| \sum\limits_{i=-s}^{s} NDVI^C(x+i,y+i,t_{1,k}) - \sum\limits_{i=-s}^{s} NDVI^C(x+i,y+i,t_2) \right|} \tag{3.4}$$

This weighting procedure assigns greater weight to the prediction if smaller changes occur between the prediction date and the observation date based on coarse data (Zhu et al., 2010). It should be noted that this weight is also calculated within the spatial moving window.

3.2.2 Test Experiment

NDVI-LMGM has been tested on three different study regions with contrasting spatial and temporal variations. Data of two study regions, Coleambally Irrigation Area and the Lower Gwydir Catchment, are provided by Emelyanova et al. (2013) as benchmark datasets for evaluating spatiotemporal data fusion methods. The Coleambally Irrigation Area (referred to as "Coleambally" hereafter), located in southern New South Wales (34.6°S, 145.6°E, 1720 × 2040 ETM+ pixels), has a total of 17 cloud free pairs of Landsat-7 ETM + and MODIS reflectance data (MOD09GA) covering the growing season from October 2001 to May 2002. The Coleambally area was selected because of its spatially heterogeneous landscape. The Lower Gwydir Catchment (referred to as "Gwydir" hereafter), located in northern New South Wales (28.9°S, 149.3°E, 3200 × 2720 ETM+ pixels), has a total of 14 cloud-free pairs of Landsat-5 TM and MODIS reflectance data (MOD09GA), from April 2004 to April 2005. Notably, a large flood struck Gwydir in December 2004, which inundated nearly 44% of the entire area. The Coleambally is considered as the spatial various region since the natural vegetation and irrigated cropland could show notable differences during different seasons. The Gwydir is considered as the temporally heterogeneous region due to the flood that occurred in December 2004. The regional NDVI dynamics for both sites are shown in Figure 3.4 to demonstrate the spatial and temporal characteristics of these benchmark data. The irrigated region in Coleambally and flood inundated region in Gwydir have also been extracted from the whole area for further comparison (see dash lines in Figure 3.4). In addition, a third area, located at Xi'an, China, is used to demonstrate the robustness of NDVI-LMGM for long term prediction. More details of these study areas could be found in (Emelyanova et al., 2013; Rao et al., 2015).

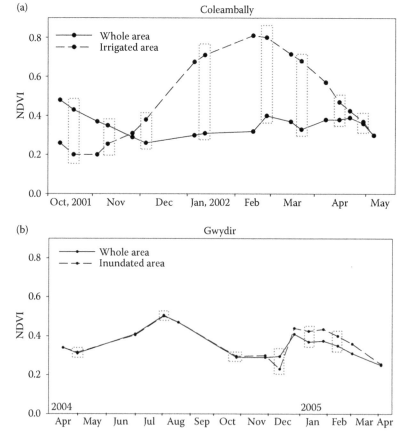

FIGURE 3.4

Dynamics of regional average NDVI for Coleambally (a) and Gwydir (b). Dates marked by dashed boxes are prediction dates used in later analysis.

3.2.2.1 Assessment over Spatial and Temporal Contrasting Regions

The performance of NDVI-LMGM prediction has been compared against STARFM (Gao et al., 2006), ESTARFM (Zhu et al., 2010), and Weighted Linear Mixing (WLM, (Busetto et al., 2008)) methods. To quantitatively compare results from these four methods, average difference (AD), average absolute difference (AAD), and average absolute relative difference (AARD) are used here.

Figure 3.5 presents the comparison of prediction results using different spatiotemporal data fusion methods. For the spatial heterogeneous site Coleambally (Figure 3.5a and b), NDVI-LMGM consistently outperforms the other three methods with the lowest AD and AAD for all prediction dates, followed by STARFM and ESTARFM, while WLM has the largest error for this study area. For the temporal contrasting region Gwydir (Figure 3.5c and d), the comparison results show a similar pattern in that predictions of NDVI-LMGM have lowest ADs and AADs across all prediction dates, followed by STARFM, ESTARFM, and WLM in this order. This result demonstrates

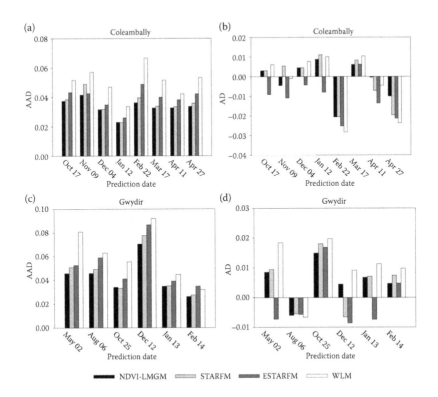

FIGURE 3.5
Accuracy comparison for prediction at the Coleambally and Gwydir sites for all prediction dates: AAD (a) and AD (b) for Coleambally, AAD (c) and AD (d) for Gwydir.

the robustness and effectiveness of NDVI-LMGM over both spatially and temporally heterogeneous landscapes.

Furthermore, prediction results for regions with the largest spatiotemporal changes in both regions, namely the irrigated rice paddy in Coleambally and flood inundated area in Gwydir, are also compared (Figure 3.6). For the irrigated rice paddy in Coleambally (Figure 3.6a–d), October 17, 2001, and February 22, 2002, are chosen for comparison, because these two dates are at the start and peak of the local crop growth season with contrasting NDVI values (0.2 for earlier date and 0.8 for later date). Although the AAD values for the two prediction dates are comparable, all four methods achieved relatively better performance for the peak of the growth season (AARD < 3%, Figure 3.6d) compared to the performance for the start of the growth season (AARD > 10%, Figure 3.6b). Moreover, NDVI-LMGM has the best performance on both dates with the smallest AAD and AARD.

For the inundated region in Gwydir, prediction results for August 6, 2004, and December 12, 2004, are compared, which are dates before and during the flood inundation, respectively. Generally, the predictions for the date with flood inundation have worse performances than predictions for the earlier date (Figure 3.6e–h). This is mainly because all methods assume the land cover type does not change, which is not valid in this inundation case. Nevertheless, NDVI-LMGM kept its superior performance compared to STARFM, ESTARFM, and WLM, with the lowest AAD and AARD. These results suggest that the relative performance of the four methods can be deteriorated for regions or seasons with smaller NDVI values, but

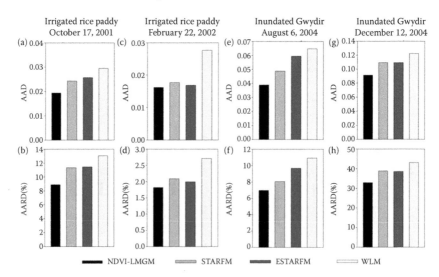

FIGURE 3.6
Accuracy assessment (AAD and AARD) for irrigated area of Coleambally on October 17, 2001 (a,b) and February 22, 2002 (c,d); inundated area of Gwydir on August 6, 2004 (e,f) and December 12, 2004 (g,h).

NDVI-LMGM still outperforms other methods for cases of seasonal changes or rapid changes.

3.2.2.2 Assessment for Long Term Prediction

The NDVI-LMGM has also been used to predict a year long NDVI time series for the third study area using a 16-day composite MODIS NDVI product (MOD13) of the year 2009 and one Landsat image acquired on June 13, 2009. This experiment is trying to evaluate the effectiveness and robustness for long term prediction. Figure 3.7 shows three examples of prediction results for both pure and mixed MODIS pixels. For the pure forest pixel (Figure 3.7a), the predicted fine pixel NDVI time series maintains the temporal dynamics as well as the magnitude of NDVI throughout the year (Figure 3.7d). For mixed MODIS pixels (Figure 3.7b,c), the NDVI-LMGM successfully separates the mixed temporal change information and magnitude of NDVI for different land cover types. For mixed MODIS pixels, predicted time series for the forest fine pixel maintains the growth information from MODIS data. On the contrary, the predicted time series for the water fine pixels and double season cropland fine pixels have reasonable NDVI seasonal dynamics and magnitude, which could be very beneficial for further applications including land cover classification, change detection, and phenology monitoring.

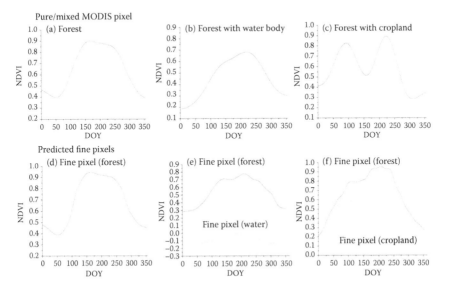

FIGURE 3.7

Examples of predicted fine-resolution time series for pure and mixed MODIS pixels: (a) pure forest MODIS time series, mixed MODIS time series with (b) forest and water, (c) forest and double season cropland; predicted fine pixel time series for (d) forest located within MODIS pixel of (a), (e) for forest and water fine pixels located within MODIS pixel of (b), (f) for forest and double season fine pixels located within MODIS pixel of (c).

3.3 Flexible Spatiotemporal Data Fusion Method (FSDAF)

3.3.1 Description of FSDAF

The input data for FSDAF include one pair of coarse- and fine-resolution images at observation date t_1 and a coarse resolution image at prediction date t_2. Before the implementation of FSDAF, input coarse- and fine-resolution data should be coregistered to the same projection and converted to the same physical quantity. FSDAF includes six main steps (Figure 3.8). Detailed descriptions of each step in FSDAF are given below.

> *Step 1. Estimate land cover fractions at t_1:* Like NDVI-LMGM, to keep the process automatic, an unsupervised classifier, ISODATA, is used in FSDAF to classify the fine-resolution image using all its spectral bands. Supervised classifiers can be also used if ground reference data are available. Additionally, available high-resolution land cover maps from other sources can be used to replace this classification step. The classification result is then used to estimate the fraction of each class in a coarse pixel, for example, $f_c(x, y)$, fraction of class c within a coarse pixel (x, y).

> *Step 2. Estimate the temporal change of each class:* Based on spectral linear mixing theory, the temporal change of a coarse pixel (x, y) for band b is the weighted sum of the temporal change of all classes within it:

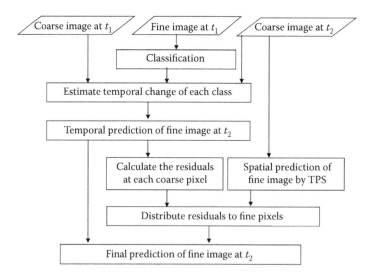

FIGURE 3.8
Flowchart of the Flexible Spatiotemporal Data Fusion method (FSDAF).

$$\Delta C(x,y,b) = \sum_{c=1}^{l} f_c(x,y) \times \Delta F(c,b) \tag{3.5}$$

where $\Delta C(x, y, b)$ is the temporal change of coarse pixel (x, y), l is the number of classes, and $\Delta F(c, b)$ is the temporal change of c-th class for band b. Equation 3.5 is valid only when no land cover type change happens between t_1 and t_2. Assuming the temporal change of each class is the same for all coarse pixels, a system of linear mixture equations could be composed by selecting n ($n > l$) coarse pixels (Equation 3.6):

$$\begin{bmatrix} \Delta C(x_1,y_1,b) \\ \vdots \\ \Delta C(x_j,y_j,b) \\ \vdots \\ \Delta C(x_n,y_n,b) \end{bmatrix} = \begin{bmatrix} f_1(x_1,y_1) & f_2(x_1,y_1) & \cdots & f_l(x_1,y_1) \\ \vdots & \vdots & & \vdots \\ f_1(x_j,y_j) & f_2(x_j,y_j) & \cdots & f_l(x_j,y_j) \\ \vdots & \vdots & & \vdots \\ f_1(x_n,y_n) & f_2(x_n,y_n) & \cdots & f_l(x_n,y_n) \end{bmatrix} \begin{bmatrix} \Delta F(1,b) \\ \vdots \\ \Delta F(c,b) \\ \vdots \\ \Delta F(l,b) \end{bmatrix} \tag{3.6}$$

$\Delta F(c, b)$ can then be solved through the inversion of Equation 3.6 using a least square method. However, collinearity and land cover type change might influence the inversion accuracy. To mitigate the influences of these issues, for each class, k "purest" coarse pixels with the highest fraction are selected. Then, among these k purest coarse pixels, ΔC of coarse pixels with abrupt land cover type change would be outliers because land cover type change often happens in a relatively small portion of the whole image and pixels with the largest changes are relatively rare. Accordingly, of the k purest coarse pixels of each class, the ones with ΔC outside of the range of 0.1~0.9 quantiles (or a narrower range, for example, 0.25~0.75, if land cover type change is large through inspecting the two coarse-resolution images) are excluded.

Step 3. Obtain temporal prediction of fine image at t_1 and estimate its residuals: After the estimation of $\Delta F(c, b)$, the reflectance of fine pixels at t_2 could be predicted by adding temporal change of each class to values at t_1 based on the classification result,

$$F_2^{TP}(x_j,y_j,b) = F_1(x_j,y_j,b) + \Delta F(c,b) \text{ if } (x_j,y_j) \text{ belongs to class } c \tag{3.7}$$

where $F_2^{TP}(x_j,y_j,b)$ is referred to as the temporal prediction of j-th fine pixels within coarse pixel (x, y). This is only valid when there is no land cover type change between t_1 and t_2. Equation 3.7 also ignores the within class variability of each class. These issues could lead to notable residuals given in Equation 3.8:

$$R(x_i, y_i, b) = \Delta C(x_i, y_i, b) - \frac{1}{m}\left[\sum_{j=1}^{m} F_2^{TP}(x_{ij}, y_{ij}, b) - \sum_{j=1}^{m} F_1(x_{ij}, y_{ij}, b)\right]. \quad (3.8)$$

To further improve the accuracy of prediction at fine resolution, the residual should be distributed to fine pixels to account for local variability and land cover type changes between t_1 and t_2.

Step 4. Get spatial prediction for guiding residual distribution: To help distribute temporal prediction residuals, FSDAF uses the thin plate spline (TPS) method to downscale information of land cover type change and local variability at prediction date t_2 contained in coarse resolution data to fine resolution. TPS is a spatial interpolation technique for point data based on spatial dependence (Dubrule, 1984). Assigning values of coarse pixel to the center point of each pixel, a spatial dependence function could be fitted using these "point-like" data by minimizing an energy function (Zhu et al., 2016). The fitted spatial dependence function would be applied to the location of each fine pixel to get a prediction at the prediction date t_2, that is, $F_2^{SP}(x_j, y_j, b)$. This spatial prediction preserves the information of land cover type change and local variability, but it misses spatial details of small objects.

Step 5. Distribute residuals to fine pixels: Errors of temporal prediction are mainly caused by land cover type change and within class variability across the image. Therefore, FSDAF distributes more residual to fine pixels with larger errors to improve the accuracy of the temporal prediction. For a homogenous landscape, the error of temporal prediction is estimated using spatial prediction at t_2 as true values,

$$E_{ho}(x_j, y_j, b) = F_2^{SP}(x_j, y_j, b) - F_2^{TP}(x_j, y_j, b). \quad (3.9)$$

The $E_{ho}(x_j, y_j, b)$ is further adjusted to 0 if its sign is different from $R(x, y, b)$. For example, a positive residual $R(x, y, b)$ means that fine pixels within one coarse pixel are more likely underestimated by the temporal prediction, so it is reasonable to only attribute errors to fine pixels with underestimation detected by Equation 3.9, and not attribute errors to any fine pixels with overestimation, and vice versa.For a heterogeneous landscape, the error for temporal prediction is assumed to be the same within one coarse pixel since no addition information is available using spatial prediction over a heterogeneous landscape.

$$E_{he}(x_j, y_j, b) = R(x, y, b). \quad (3.10)$$

The homogeneity of any fine pixel is characterized by a homogeneity index calculated using a moving window with the size of a coarse pixel:

$$HI(x_j, y_j) = \frac{\left(\sum_{k=1}^{N} I_k\right)}{N} \tag{3.11}$$

where $I_k = 1$ when the k-th fine pixel within the moving window belongs to the same class as the central fine pixel (x_j, y_j), otherwise $I_k = 0$. The final error estimation for the fine pixel (x_j, y_j) is found by summing these two cases through HI:

$$CW(x_i, y_j, b) = E_{ho}(x_j, y_j, b) \times HI(x_j, y_j) + E_{he}(x_j, y_j, b) \times [1 - HI(x_j, y_j)]. \tag{3.12}$$

The weight of distributing residual to each fine pixel is then calculated as:

$$W(x_j, y_j, b) = \frac{CW(x_j, y_j, b)}{\sum_{j=1}^{m} CW(x_j, y_j, b)}. \tag{3.13}$$

Then, the residual distributed to j-th fine pixel is:

$$r(x_j, y_j, b) = N \times R(x, y, b) \times W(x_j, y_j, b). \tag{3.14}$$

The final prediction of the total change of a fine pixel between t_1 and t_2 is:

$$\Delta F(x_j, y_j, b) = r(x_j, y_j, b) + \Delta F(c, b) \text{ if } (x_j, y_j) \text{ belongs to class } c. \tag{3.15}$$

Step 6. Obtain a robust prediction of fine image using neighborhood: The previous steps are all implemented within each coarse pixel, which could cause a block effect similar to other unmixing based methods. FSDAF uses additional neighborhood information to get more robust prediction of fine pixel values at t_2 similar to STARFM and ESTARFM (Gao et al., 2006; Zhu et al., 2010). For each fine pixel, n similar fine pixels with the smallest spectral differences from the target fine pixel are selected within its neighborhood. The spectral difference between k-th fine pixel and the target pixel (x_j, y_j) is defined as:

$$S_k = \sum_{b=1}^{B} [|F_1(x_k, y_k, b) - F_1(x_j, y_j, b)| / F_1(x_j, y_j, b)]. \tag{3.16}$$

Based on trial-and-error experiment, Zhu et al. (2016) recommends 20 similar pixels to balance robustness and efficiency since more similar pixels require more computation time. After similar pixels are selected, the weight of each similar pixel is determined by the spatial distances between similar pixels and the target pixel.

$$w_k = \frac{(1/D_k)}{\sum_{k=1}^{n}(1/D_k)},$$

where

$$D_k = 1 + \sqrt{(x_k - x_i)^2 + (y_k - y_i)^2}/(w/2) \qquad (3.17)$$

where w is the size of the neighborhood, which is manually determined by users based on the homogeneity of the study area and is commonly the size of one to three coarse pixels.

Change values of all similar pixels are then summed by weights to get the final change value of the target fine pixel. Adding this final change estimation to the observation at t_1 yields a more reliable final prediction of the target pixel at t_2:

$$\hat{F}_2(x_j, y_j, b) = F_1(x_j, y_j, b) + \sum_{k=1}^{n} w_k \times \Delta F(x_k, y_k, b). \qquad (3.18)$$

3.3.2 Test Experiment

FSDAF has been tested against both simulated data and real MODIS/Landsat data pairs. The simulated data (Figure 3.9a–d) have two pairs of coarse- and fine-resolution images. The simulated fine images have notable land cover type change, that is, the enlargement of the circle object from Figure 3.9b–d. The real MODIS/Landsat data pairs are extracted from the benchmark dataset provided by Emelyanova et al. (2013), which has also been used for evaluating NDVI-LMGM. The study area, located in northern New South Wales, Australia (149.2815°E, 29.0855°S), covers approximately a region of 20 km × 20 km with large parcels of croplands and natural vegetation. Two pairs of MODIS/Landsat images were acquired on November 26th and December 12th, 2002. Between these two acquisition dates, a large flood occurred within the study area, causing substantial land cover type changes (inundation) (Figure 3.10a–d) (Emelyanova et al., 2013). To test the FSDAF, the data pairs of fine/coarse images of the earlier date and the coarse image of the later date are used as inputs to predict the fine image of the later date, which will be compared to the observed fine image in Figure 3.10d.

The results produced by FSDAF have been compared with other two spatiotemporal fusion methods, that is, STARFM (Gao et al., 2006) and Unmixing-Based Data Fusion (UBDF, (Zurita-Milla et al., 2008)). Different

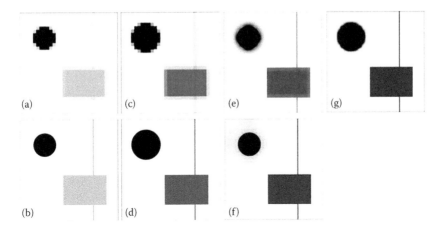

FIGURE 3.9
Simulated coarse (a,c) and fine (b,d) images at t_1 and t_2; predicted fine image at t_2 by STARFM (e), UBDF (f), and FSDAF (g) using (a–c) as inputs for each method.

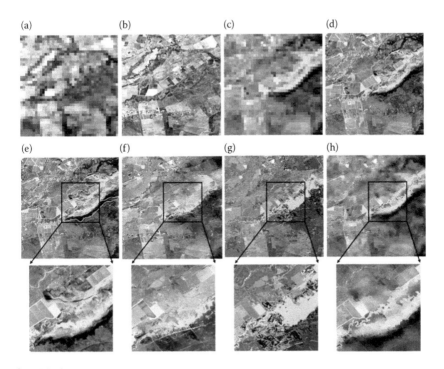

FIGURE 3.10
(See color insert.) (a) MODIS image of November 26th; (b) Landsat image of November 26th; (c) MODIS image of December 12th; (d,e) Landsat image of December 12th; predicted images of December 12th by STARFM (f), UBDF (g), and FSDAF (h). White line in (e) delineates the boundaries of inundated area.

indices are used to quantitatively assess predictions, including root mean square error (RMSE), average difference (AD), correlation coefficient (r), and structure similarity (SSIM) (Wang et al., 2004).

3.3.2.1 Assessment of Simulated Results

Figure 3.9e–g shows the comparison among prediction results using STARFM, UBDF, and FSDAF. For the expanded circle object, FSDAF predicts more similar results compared to the real image, while UBDF produces a blurred zone around the circle. Although STARFM generates less blurry prediction compared to UBDF, it is still quite different compared to the true image. The quantitative comparisons among three different methods are summarized in Table 3.1. Among three spatiotemporal data fusion methods, FSDAF has the best performance in terms of all indices, following by STARFM and UBDF. In addition, both FSDAF and UBDF could produce nearly unbiased predictions while STARFM has a notable positive bias compared to the true image.

3.3.2.2 Assessment of Fusing Real Satellite Images

Figure 3.10e–h presents Landsat-like images predicted by the three methods, the original Landsat image of December 12, 2004, and zoom-in scenes of the inundated regions to show more details. Prediction of FSDAF shows better agreement with the true Landsat image, while STARFM's prediction is more "blurry" and UBDF's has substantial errors for the inundated area. Table 3.2 summarizes the quantitative assessment for predictions using three spatiotemporal fusion methods. For the whole study area, FSDAF provides the best predictions for all six bands with the smallest RMSE and highest r and SSIM. Although STARFM has lower accuracy compared to FSDAF, its performance is much better than UBDF. The superior performance of FSDAF is confirmed by the scatter plot between true near infrared Landsat reflectance and three predictions (Figure 3.11). For the inundated region (marked zone in Figure 3.10e), FSDAF has the consistently best prediction for all six bands, while both STARFM and UBDF show notably larger prediction errors compared with the results of the whole region (Table 3.2). The 5th band is

TABLE 3.1

Accuracy Assessment of Three Data Fusion Methods Applied to the Simulated Dataset in Figure 3.9

Method	RMSE	r	AD	SSIM
STARFM	0.0405	0.962	0.0020	0.9592
UBDF	0.0583	0.919	0.0003	0.9131
FSDAF	0.0256	0.984	0.0001	0.9843

Note: The units are reflectance (RMSE = Root Mean Square Error, r = correlation coefficient, AD = average difference, SSIM = structural similarity).

TABLE 3.2

Quantitative Assessment of Three Data Fusion Methods over the Whole Study Region and Inundated Regions Only

		STARFM				UBDF				FSDAF			
		RMSE	r	AD	SSIM	RMSE	r	AD	SSIM	RMSE	r	AD	SSIM
Whole region	B1	0.011	0.82	0.000	0.803	0.018	0.51	0.000	0.505	0.010	0.86	0.000	0.848
	B2	0.016	0.81	0.000	0.800	0.025	0.55	0.000	0.551	0.013	0.87	0.000	0.857
	B3	0.019	0.79	0.000	0.778	0.030	0.52	0.000	0.523	0.016	0.85	0.000	0.843
	B4	0.026	0.88	0.000	0.868	0.045	0.68	0.000	0.679	0.022	0.92	0.000	0.910
	B5	0.045	0.84	0.000	0.828	0.071	0.61	0.001	0.610	0.040	0.88	0.001	0.872
	B7	0.033	0.84	0.000	0.827	0.052	0.61	0.001	0.604	0.030	0.87	0.001	0.864
Inundated region	B1	0.004	0.66	0.004	0.596	0.006	0.37	0.005	0.366	0.003	0.87	0.000	0.857
	B2	0.006	0.62	0.005	0.572	0.008	0.38	0.007	0.382	0.005	0.88	0.000	0.859
	B3	0.007	0.59	0.006	0.544	0.010	0.35	0.008	0.345	0.006	0.87	0.000	0.858
	B4	0.007	0.69	0.008	0.659	0.012	0.50	0.014	0.474	0.005	0.87	0.002	0.854
	B5	0.013	0.60	0.014	0.563	0.019	0.32	0.022	0.303	0.010	0.81	0.005	0.787
	B7	0.008	0.64	0.009	0.600	0.013	0.32	0.015	0.310	0.007	0.80	0.003	0.779

Note: RMSE = Root Mean Square Error, r = correlation coefficient, AD = average difference, SSIM = structural similarity.

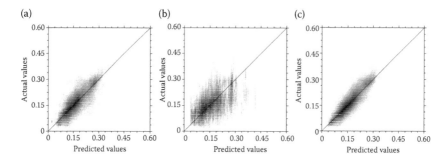

FIGURE 3.11
Scatter plots of the actual and predicted values (a) STARFM, (b) UBDF, (c) FSDAF for near
infrared band (darker color indicates a higher density of points, and the line is 1:1 line).

the one with the most change when flooded. For this band, the $SSIM$ values
of FSDAF, STARFM, and UBDF are 0.787, 0.563, and 0.303, respectively,
suggesting that FSDAF is more powerful for predicting pixels located in a
complex area, that is, a transitional area of change versus non-change, which
is a condition that challenges existing spatiotemporal data fusion methods.

3.4 Conclusions and Discussion

Time series of remote sensing data with high spatial resolution and frequent
coverage is essential for monitoring land surface dynamics, especially for
heterogeneous landscapes. Unfortunately, no single sensor could provide this
type of data currently due to the trade-off between spatial resolution and
acquisition frequency (Gao et al., 2010). Spatiotemporal data fusion methods
have been developed during the last two decades to combine multi-sensor
remote sensing data with different spatial and temporal characteristics.
The synthetic data generated by spatiotemporal data fusion methods are
expected to convey more spatial details and temporal change information
than single sensor data alone (Gao et al., 2015). In this chapter, we introduced
two advanced spatiotemporal fusion methods, that is, NDVI-LMGM (Rao
et al., 2015; Zhu et al., 2016) and FSDAF (Zhu et al., 2016). FSDAF is designed
to predict reflectance images, while NDVI-LMGM is designed to generate
synthetic NDVI time series.

The basic assumption of FSDAF and NDVI-LMGM is similar in that the
temporal change of coarse pixels is the weighted average of temporal changes
of each class at fine-resolution using class fraction as weights. Under this
basic assumption, NDVI-LMGM and FSDAF adopt different strategies to
better estimate the real temporal change. NDVI-LMGM assumes that the
temporal change of each class is the same at local regions rather than the
whole image. Therefore, NDVI-LMGM only uses coarse pixels within a

small spatial moving window to unmix temporal changes to fine resolution. This assumption takes the within class variability into consideration, which makes it outperform other spatiotemporal data fusion methods. To ensure the unmixed temporal changes of each class are a reasonable value, NDVI-LMGM uses a constrained least square method instead of a regular least square method. Results with different benchmark datasets demonstrate the effectiveness and robustness of NDVI-LMGM. FSDAF realizes the temporal changes for each class through solving the mixing equations has residuals due to within class variability and local land cover type changes. To account for the local land cover changes, FSDAF uses thin plate spline interpolation to help distribute the residual. In this manner, FSDAF could help produce robust and accurate predictions of high resolution images even when abrupt land cover changes happen during the data fusion period.

Due to different purposes and implementation requirements, FSDAF and NDVI-LMGM have been evaluated separately in this chapter. However, they have both been evaluated using the benchmark data of Australia with a notable flooding event during the study period (i.e., December 2004). NDVI-LMGM has demonstrated consistent superior performances over existing unmixing based methods as well as weighted function based methods in the whole study region (Figure 3.5). However, the performance of NDVI-LMGM deteriorates when it is evaluated for the inundated regions only, but still out performs other fusion methods. The deterioration of performance is partially linked to the violation of the assumption that there are no local land cover changes (Figure 3.6). When using FSDAF to fuse reflectance data, it shares a similar performance with the weighted function based reflectance fusion method (i.e., STARFM), but outperforms the existing unmixing based reflectance fusion method (i.e., UBDF) for the whole study region. Moreover, when fused reflectance is evaluated only for the inundated area, FSDAF stands out with near zero biases and low uncertainty (Table 3.2). All other data fusion methods (including NDVI-LMGM) assumes there are no local abrupt surface type changes, which is not valid for the inundation. The violation of this assumption creates a large uncertainty for data fusion, while FSDAF is robust to the local abrupt surface type change due to the thin-plate-interpolation step.

Even though most spatiotemporal data fusion methods are developed for reflectance or vegetation indices because of their popularity, they can also be used to fuse other reflectance based indices and products if they are linearly additive in space (Zhu et al., 2016). To produce fused reflectance based indices or products, there are two strategies: (1) the direct strategy which directly applies spatiotemporal data fusion methods to coarse- and fine-resolution products to generate synthetic products; (2) the indirect strategy which produces fused multispectral images first and generates synthetic products using fused images. A recent study has shown that if the final objective is to generate synthetic products, the direct approach is preferred no matter which spatiotemporal data fusion methods are used (Jarihani et al., 2014). Compared to the indirect strategy, the direct fusion strategy could reduce computational cost and avoid error propagation from different spectral bands (for more details, please refer to Jarihani et al., 2014).

Most spatiotemporal data fusion methods, including methods introduced in this chapter, are tested for fusing Landsat and MODIS data. However, these methods could also be extended to other similar data when fine- and coarse-resolution data have similar characteristics (i.e., reflectance of similar bands or same indices), such as data from Visible and Infrared Imaging Radiometer Suite (VIIRS) on board Suomi National Polar-orbiting Partnership (S-NPP) and future Joint Polar Satellite System (JPSS) missions. Even though spatiotemporal data fusion methods demonstrate their potential for providing synthetic data with high resolution and frequent coverage, these data cannot replace the actual data from satellites, especially when the change over land surface is abrupt and subtle. It has been suggested by experiments in the previous sections that most data fusion results are worse for heterogeneous landscapes or when the land cover types have notable changes. Additionally, when the change over a land surface is missed (due to cloud) or too small to be captured by the coarse resolution data, this information is very difficult to be reconstructed in the synthetic data unless extra fine-resolution data are available after the change occurred (Gao et al., 2015). It should also be noted that it is challenging to predict data that are distant from available observations. Therefore, more frequent fine-resolution data could help capture frequent and small surface changes that would reduce the uncertainty of the synthetic data. With more Landsat-like observations, such as multispectral images (10–20 m) acquired by MultiSpectral Instrument (MSI) onboard European Space Agency's (ESA) Sentinel-2 missions, it is possible to have daily synthetic data to monitor rapid land surface changes over heterogeneous landscape.

___

Acknowledgments

Xiaolin Zhu is supported by the Hong Kong Polytechnic University Research Grant (1-ZE6Q), the Hong Kong Research Grants Council Early Career Scheme (Grant No. F-PP47), and the National Natural Science Foundation of China (Project No.41701378). Yuhan Rao is supported by China Scholarship Council.

___

References

Amorós-López, J., Gómez-Chova, L., Alonso, L., Guanter, L., Zurita-Milla, R., Moreno, J., Camps-Valls, G. 2013. Multitemporal fusion of Landsat/TM and ENVISAT/MERIS for crop monitoring. *Int. J. Appl. Earth Obs. Geoinformation* 23, 132–141. doi:10.1016/j.jag.2012.12.004

Busetto, L., Meroni, M., Colombo, R. 2008. Combining medium and coarse spatial resolution satellite data to improve the estimation of sub-pixel NDVI time series. *Remote Sens. Environ.* 112, 118–131. doi:10.1016/j.rse.2007.04.004

Chen, J., Chen, J., Liao, A., Cao, X., Chen, L., Chen, X., He, C., Han, G., Peng, S., Lu, M., Zhang, W., Tong, X., Mills, J. 2015. Global land cover mapping at 30 m resolution: A POK-based operational approach. *ISPRS J. Photogramm. Remote Sens., Glob. Land Cover Mapp. Monit.* 103, 7–27. doi:10.1016/j.isprsjprs.2014.09.002

Chen, J., Jönsson, P., Tamura, M., Gu, Z., Matsushita, B., Eklundh, L. 2004. A simple method for reconstructing a high-quality NDVI time-series data set based on the Savitzky–Golay filter. *Remote Sens. Environ.* 91, 332–344. doi:10.1016/j.rse.2004.03.014

Chen, X., Yang, D., Chen, J., Cao, X. 2015. An improved automated land cover updating approach by integrating with downscaled NDVI time series data. *Remote Sens. Lett.* 6, 29–38. doi:10.1080/2150704X.2014.998793

Dubrule, O. 1984. Comparing splines and kriging. *Comput. Geosci.* 10, 327–338. doi:10.1016/0098-3004(84)90030-X

Emelyanova, I.V., McVicar, T.R., Van Niel, T.G., Li, L.T., van Dijk, A.I.J.M. 2013. Assessing the accuracy of blending Landsat–MODIS surface reflectances in two landscapes with contrasting spatial and temporal dynamics: A framework for algorithm selection. *Remote Sens. Environ.* 133, 193–209. doi:10.1016/j.rse.2013.02.007

Gao, F., Hilker, T., Zhu, X., Anderson, M., Masek, J., Wang, P., Yang, Y. 2015. Fusing Landsat and MODIS Data for Vegetation Monitoring. *IEEE Geosci. Remote Sens. Mag.* 3, 47–60. doi:10.1109/MGRS.2015.2434351

Gao, F., Masek, J., Schwaller, M., Hall, F. 2006. On the blending of the landsat and MODIS surface reflectance: Predicting daily Landsat surface reflectance. *IEEE Trans. Geosci. Remote Sens.* 44, 2207–2218. doi:10.1109/TGRS.2006.872081

Gao, F., Masek, J., Wolfe, R.E. 2009. Automated registration and orthorectification package for landsat and landsat-like data processing. *J. Appl. Remote Sens.* 3, 33515–33520. doi:10.1117/1.3104620

Gao, F., Masek, J.G., Wolfe, R.E., Huang, C. 2010. Building a consistent medium resolution satellite data set using moderate resolution imaging spectroradiometer products as reference. *J. Appl. Remote Sens.* 4, 043526. doi:10.1117/1.3430002

Gevaert, C.M., García-Haro, F.J. 2015. A comparison of STARFM and an unmixing-based algorithm for Landsat and MODIS data fusion. *Remote Sens. Environ.* 156, 34–44. doi:10.1016/j.rse.2014.09.012

Guerschman, J.P., Paruelo, J.M., Bella, C.D., Giallorenzi, M.C., Pacin, F. 2003. Land cover classification in the Argentine Pampas using multi-temporal Landsat TM data. *Int. J. Remote Sens.* 24, 3381–3402. doi:10.1080/0143116021000021288

Hilker, T., Wulder, M.A., Coops, N.C., Linke, J., McDermid, G., Masek, J.G., Gao, F., White, J.C. 2009. A new data fusion model for high spatial- and temporal-resolution mapping of forest disturbance based on Landsat and MODIS. *Remote Sens. Environ.* 113, 1613–1627. doi:10.1016/j.rse.2009.03.007

Huang, B., Song, H. 2012. Spatiotemporal reflectance fusion via sparse representation. *IEEE Trans. Geosci. Remote Sens.* 50, 3707–3716. doi:10.1109/TGRS.2012.2186638

Jarihani, A.A., McVicar, T.R., Van Niel, T.G., Emelyanova, I.V., Callow, J.N., Johansen, K. 2014. Blending landsat and MODIS data to generate multispectral indices: A comparison of "Index-then-Blend" and "Blend-then-Index" approaches. *Remote Sens.* 6, 9213–9238. doi:10.3390/rs6109213

Ju, J., Roy, D.P. 2008. The availability of cloud-free Landsat ETM+ data over the conterminous United States and globally. *Remote Sens. Environ.* 112, 1196–1211. doi:10.1016/j.rse.2007.08.011

Lu, M., Chen, J., Tang, H., Rao, Y., Yang, P., Wu, W. 2016. Land cover change detection by integrating object-based data blending model of Landsat and MODIS. *Remote Sens. Environ.* 184, 374–386. doi:10.1016/j.rse.2016.07.028

Rao, Y., Zhu, X., Chen, J., Wang, J. 2015. An improved method for producing high spatial-resolution NDVI time series datasets with multi-temporal MODIS NDVI data and landsat TM/ETM+ images. *Remote Sens.* 7, 7865–7891. doi:10.3390/rs70607865

Shen, M., Tang, Y., Chen, J., Zhu, X., Zheng, Y. 2011. Influences of temperature and precipitation before the growing season on spring phenology in grasslands of the central and eastern Qinghai-Tibetan Plateau. *Agric. For. Meteorol.* 151, 1711–1722. doi:10.1016/j.agrformet.2011.07.003

Song, H., Huang, B. 2013. Spatiotemporal satellite image fusion through one-pair image learning. *IEEE Trans. Geosci. Remote Sens.* 51, 1883–1896. doi:10.1109/TGRS.2012.2213095

Wang, Z., Bovik, A.C., Sheikh, H.R., Simoncelli, E.P. 2004. Image quality assessment: From error visibility to structural similarity. *IEEE Trans. Image Process.* 13, 600–612. doi:10.1109/TIP.2003.819861

Zhu, X., Chen, J., Gao, F., Chen, X., Masek, J.G. 2010. An enhanced spatial and temporal adaptive reflectance fusion model for complex heterogeneous regions. *Remote Sens. Environ.* 114, 2610–2623. doi:10.1016/j.rse.2010.05.032

Zhu, X., Helmer, E.H., Gao, F., Liu, D., Chen, J., Lefsky, M.A. 2016. A flexible spatiotemporal method for fusing satellite images with different resolutions. *Remote Sens. Environ.* 172, 165–177. doi:10.1016/j.rse.2015.11.016

Zhu, X., Liu, D., Chen, J. 2012. A new geostatistical approach for filling gaps in Landsat ETM+ SLC-off images. *Remote Sens. Environ.* 124, 49–60. doi:10.1016/j.rse.2012.04.019

Zhukov, B., Oertel, D., Lanzl, F., Reinhackel, G. 1999. Unmixing-based multisensor multiresolution image fusion. *IEEE Trans. Geosci. Remote Sens.* 37, 1212–1226. doi:10.1109/36.763276

Zurita-Milla, R., Clevers, J.G.P.W., Schaepman, M.E. 2008. Unmixing-based landsat TM and MERIS FR data fusion. *IEEE Geosci. Remote Sens. Lett.* 5, 453–457. doi:10.1109/LGRS.2008.919685

Part II

Feature Development and Information Extraction

4

Phenological Inference from Times Series Remote Sensing Data

Iryna Dronova and Lu Liang

CONTENTS

4.1   Introduction 69
4.2   Single-Season Phenological Analyses 70
    4.2.1   Spectral Indicators of Phenology 70
    4.2.2   Basic Seasonal Phenological Trajectory 71
    4.2.3   Phenological Variation Represented by Seasonal Trajectories 72
4.3   Common Applications of Single-Year Phenology 74
    4.3.1   Agricultural Mapping and Monitoring of Crops 74
    4.3.2   Forest Mapping and Ecosystem Analyses 74
    4.3.3   Hydro-Phenological Analyses of Complex Flooded
             Landscapes 76
4.4   Multi-Year Phenological Inference 77
    4.4.1   Local-Scale: Phenocam Observations 77
    4.4.2   Regional Analyses of Greenness Trends 78
          4.4.2.1   Basic Trend Analyses 78
          4.4.2.2   Trajectory-Based Landscape Change Analyses 79
          4.4.2.3   Continuous Change Detection and Classification
                   of Land Cover 81
    4.4.3   Broad-Scale Phenological Analyses with Multi-Year
          Seasonal Data 81
          4.4.3.1   Multi-Year Inference with Phenological Curves 81
          4.4.3.2   Percent above Threshold Approaches 82
4.5   Applications and the Importance of Ancillary Factors 82
References 84

4.1 Introduction

Phenology broadly refers to seasonal phenomena associated with biological cycles and activity of plants and animals triggered by changes in temperature, precipitation, daylight, and other environmental factors. While specific schedule and timing of these changes may vary from year to year due to natural variations in climate and weather, directional phenological shifts such as consistently earlier times of leaf flushing and flowering of plants may provide important early warning signals of major climatic transitions and global warming effects (Cleland et al., 2007). Such shifts may be observed not only in natural systems, but also in heavily human-dominated landscapes. For example, urban regions with higher temperatures compared to exurbia and wildlands may also display longer growing seasons (Zipper et al., 2016). The ability to monitor phenological changes is crucial for understanding the underlying physical transitions in the landscapes and dynamics of important ecosystem services such as productivity and uptake of atmospheric greenhouse gas CO_2 by plants, agricultural crop pollination, and biological pest control associated with plant-animal seasonal cycles and many other (Cho et al., 2017; Klosterman et al., 2014; Richardson et al., 2012; Toomey et al., 2015).

Time series of remote sensing data are a particularly useful source of phenological information over large regional extents, where field observations are difficult to perform with sufficient spatial and temporal coverage (Reed et al., 1994; White et al., 1997; Zhang et al., 2003). Cost-effective monitoring of landscapes via repeatedly collected images provides spatially explicit information on vegetation greenness dynamics and related ecosystem processes (Klosterman et al., 2014; Richardson et al., 2012; Toomey et al., 2015) or even seasonal patterns of animal activity and management (Hilker et al., 2014; Neuenschwander and Crews, 2008). Recent advances in very high resolution, near-surface and *in situ* remote sensing provide new cost-effective opportunities to detect more fine scale and complex phenological dynamics and thus enhance ecological monitoring and adaptive management of ecosystem functions and services.

The objective of this chapter is to review the key remote sensing-based phenological metrics, research approaches, and their major applications in ecosystem function and climate change studies. It starts by introducing the primary metrics of single-season phenological parameters based on the indicators of vegetation greenness and schedules of its dynamics. It then proceeds to discuss the inter-annual metrics and approaches to detect the longer term phenological shifts triggered by climate, land management, and other factors. The final section concludes with a review of major applications at local, regional, and continental to global scales.

4.2 Single-Season Phenological Analyses

4.2.1 Spectral Indicators of Phenology

Many remote sensing studies characterize phenology using proxies of seasonally variable plant greenness, such as spectral vegetation indices, or SVIs (several common metrics shown in Table 4.1). The choice of specific greenness indicators varies among studies and applications. For example,

TABLE 4.1

Examples of Spectral Vegetation Indices used in Phenological Remote Sensing Studies

Vegetation Index	Formula	Reference
Enhanced vegetation index (EVI)	$G \times (\rho_{NIR} - \rho_{red})/(\rho_{NIR} + C1 \times \rho_{red} - C2 \times \rho_{blue} + L)$, where $C1 = 6, C2 = 7.5, L = 1$ and $G = 2.5$	Huete et al. (2002)
Normalized difference vegetation index (NDVI)	$(\rho_{NIR} - \rho_{red})/(\rho_{NIR} + \rho_{red})$	Kriegler et al. (1969), Rouse et al. (1974)
Normalized difference water index (NDWI)	$(\rho_{green} - \rho_{NIR})/(\rho_{NIR} + \rho_{green})$	McFeeters (1996)
Green NDVI (GNDVI)	$(\rho_{NIR} - \rho_{green})/(\rho_{NIR} + \rho_{green})$	Gitelson and Merzlyak (1997)
Land surface water index (LSWI)	$(\rho_{NIR} - \rho_{SWIR})/(\rho_{NIR} + \rho_{SWIR})$	Xiao et al. (2004)
Green chromatic coordinate (GCC)[a]	$\rho_{green}/(\rho_{green} + \rho_{red} + \rho_{blue})$	Woebbecke et al. (1995)
Excess green index (ExG)[a]	$2 \times \rho_{green} - (\rho_{red} + \rho_{blue})$	Woebbecke et al. (1995)
Visible atmospherically resistant index (VARI)[a]	$(\rho_{green} - \rho_{red})/(\rho_{red} + \rho_{green})$	Sakamoto et al. (2005)
Green-red ratio (grR)[a]	ρ_{green}/ρ_{red}	Sonnentag et al. (2011)
Red-blue ratio (rbR)[a]	ρ_{red}/ρ_{blue}	Sonnentag et al. (2011)

[a] Do not require a near infrared band and thus can be quantified from regular RGB digital photos.

phenological analyses using satellite or airborne imagery over large regional extents often rely on popular SVIs such as normalized difference vegetation index (NDVI) and enhanced vegetation index (EVI), that can be estimated by a variety of datasets in visible (e.g., red, green, blue) and near infrared electromagnetic regions. Moisture-sensitive indices such as normalized difference water index (NDWI) or land surface water index (LSWI) can be helpful in areas where phenological signatures are affected by inundation or variation in plant water content, such as flooded agriculture (e.g., Dong et al., 2015, 2016). In turn, growing applications of near surface remote sensing (where infrared sensitivity may be still too costly for project budgets) have successfully utilized indices based on visible only spectral regions (Table 4.1) from simple red-green-blue (RGB) digital camera imagery.

4.2.2 Basic Seasonal Phenological Trajectory

Single-year phenology is often characterized from a spectral-temporal trajectory of a given greenness indicator (Figure 4.1) at different days of year (DOY), sometimes interpolated to higher temporal frequency using

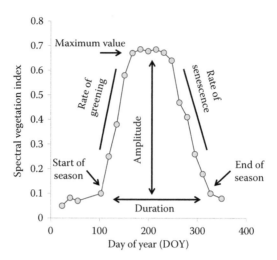

FIGURE 4.1
A generic phenological trajectory for one growing season.

mathematical functions (e.g., Zhang et al., 2006; Zhong et al., 2012). Figure 4.1 shows a generic hump-shaped NDVI trajectory typical of deciduous, annual, and seasonally harvested vegetation types, where greenness increases from the stage with no green leaves to peak biomass stage and ultimately declines during the end-of-season leaf senescence.

Several useful metrics can be estimated from such data to characterize phenology and compare it among different vegetation types, regions or years (Figure 4.1). Common *metrics of greenness* include maximum and minimum SVI values and their difference or amplitude; SVI values corresponding to start and end of the growing season; and rates of greenness and senescence indicated by the steepness in the curves during the rapid seasonal transitions. Some studies also use time-integrated greenness as the sum of all values under the seasonal curve and seasonally averaged greenness as the mean of single-season SVI values to account for temporally non-uniform leaf biomass and contributions to ecosystem processes (Dronova et al., 2011; White et al., 1997). In turn, *metrics of time* include start and end of the growing season, often represented by inflection points on each side of the curve (Figure 4.1); duration or length of the growing season as the difference between start and end dates; and the timing of maximum greenness.

4.2.3 Phenological Variation Represented by Seasonal Trajectories

Importantly, shapes and metrics of seasonal spectral trajectories may vary among vegetation types thus highlighting their unique physiological cycles and environmental adaptations. Examples in Figure 4.2 show such differences for several land cover types in California, USA's San Francisco Bay Area region using EVI data from Moderate Resolution Imaging Spectroradiometer

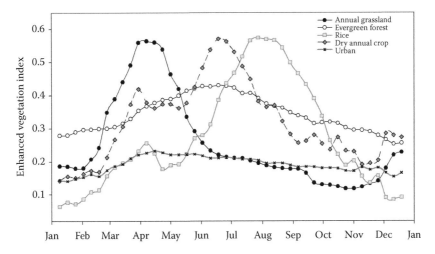

FIGURE 4.2
An example of per pixel seasonal phenological profiles of five land cover types in California, USA, based on 500-m 16-day Moderate Resolution Imaging Spectroradiometer (MODIS) Enhanced Vegetation Index (EVI) series for 2014.

(MODIS). Naturalized exotic annual grasses common in the area exhibit a unique trajectory of greenness increasing during the wet winter season and reaching a peak in April, followed by a senescent phase until the end of the subsequent fall (Figure 4.2). In contrast, annual crops often reach maximum biomass and greenness in summer; their early development stages may show temporary spikes and declines of greenness due to pruning or weed removal. In wet agriculture such as rice, short term declines in greenness may also reflect field flooding at the earlier phase of crop growth, while low winter greenness corresponds to flooded fields without green vegetation. Evergreen woodlands in Figure 4.2 show characteristically high winter EVI values compared to nonevergreen vegetation due to maintenance of green leaves; yet, their greenness still increases during spring to early summer when the new leaf cohorts are developing.

In contrast, urban areas show relatively little change and low overall greenness throughout the year (Figure 4.2) due to the contributions of impervious and other nonvegetated surfaces to pixel level signals, which may attenuate the signal from vegetated surfaces. The analyses of urban region phenologies thus need to carefully consider such subpixel mixing and scale effects to avoid interpreting low variation in greenness as the lack of seasonality in urban vegetation. Such unique signatures become instrumental in classification of complex landscapes or vegetation types using multi-seasonal remote sensing data. By including the dates where phenological contrasts among species or plant types promote distinctiveness of their spectral signals, classes may be more easily differentiated than at the peak biomass stage when they may be more similar.

4.3 Common Applications of Single-Year Phenology

4.3.1 Agricultural Mapping and Monitoring of Crops

Unique seasonal "signatures" of different vegetation types become critically important for discriminating plant species and communities that may be otherwise similar in single-date images or near the peak of their biomass and canopy cover. This issue is particularly relevant to crops, and thus phenological data become important in agricultural mapping of crop distribution and subsequent modeling of their biomass, yield, water use, evapotranspiration, and other properties (e.g., Dong et al., 2015; Sakamoto et al., 2005; Zhong et al., 2011, 2012). Agricultural landscapes often contain a variety of crop types within even relatively small regions, and discriminating them accurately requires multiple metrics described in Section "Basic Seasonal Phenological Trajectory" and even more specific indicators. For example, Zhong et al. (2011) used the points of mid-rise in the increase and descent of the greenness trajectory and the number of days between then as indicators of crop-specific growth schedules to facilitate mapping crops with similar overall schedules (e.g., almonds and pistachios), in the California, USA's San Joaquin Valley.

Because of their unique schedules of sprouting, greening, fruiting, and harvesting, many common crop types can be distinguished especially well with high temporal frequency observations. Two studies in California (Zhong et al., 2011, 2012) used MODIS-based phenological information to differentiate several common crops using MODIS and combined MODIS-Landsat time series. In the latter study area, Mediterranean-climate winter allows for a relatively high variety of crop types, sometimes with multiple rotations per year, which makes high temporal frequency crucial for their accurate identification. MODIS data have been also used to detect specific crop types such as rice paddies (Sakamoto et al., 2005), although temporally sparser Landsat datasets have also been useful due to their sensitivity to contrasts between flooded and dry stages of the rice cycle (e.g., Dong et al., 2015). Importantly, however, coarser resolution of MODIS products may become a problem when different crop types are mixed inside large pixels. A promising solution to this problem is using "temporal unmixing" approaches, such as independent components analysis, to decompose time series of coarse pixels into relative contributions of their potential endmembers derived from representative phenologies of candidate crops (e.g., Ozdogan, 2010).

4.3.2 Forest Mapping and Ecosystem Analyses

Single-season phenology of vegetation has also been very useful in remote sensing analyses of forests, where accurate detection of tree canopy composition is necessary to understand biodiversity, habitats, successional status, and important ecosystem functions (Bergen and Dronova, 2007; Dymond et al., 2002; Melaas et al., 2013). Phenological variation can be particularly informative in regions with distinct seasonality (e.g., temperate, boreal, and high-altitude forests) and the presence of both deciduous and evergreen species with contrasting seasonal trajectories. Combined uses of images from "leaf-on" and "leaf-off" stages of deciduous vegetation can help to differentiate forest types and their dominant successional tendencies based on the contrast between deciduous and conifers in the overstory and understory (e.g., Bergen and Dronova, 2007; Mickelson et al., 1998; Wolter et al., 1995). Specific metrics such as seasonal amplitudes or between-date differences in greenness indicators can effectively highlight such differences (e.g., Dymond et al., 2002; Melaas et al., 2013), as can be also seen in Figure 4.3.

Forest studies have also applied phenological information beyond simple mapping, to model ecosystem parameters, such as aboveground net primary productivity (ANPP), based on their correlations with SVIs. However, single-date productivity assessments may not accurately represent annual changes in vegetation, while repeated measurements throughout the growing season may be unfeasible over large spatial extents. Instead, time series of NDVI and other indices may be used to create seasonally averaged indicators of plant function (Dronova et al., 2011; White et al., 1997). For instance, in the analysis of ecosystem properties in a northern US temperate forest region, seasonally averaged NDVI explained ~75% variation in field measured ANPP, while single-date NDVI – only 52% (Dronova et al., 2011). An important consideration in such efforts is the choice of a time frame over which a seasonal curve is reliably representative of vegetation dynamics – that is, the time frame between leaf flushing and senescence of deciduous vegetation. Typically the start and the end of the growing season are determined by

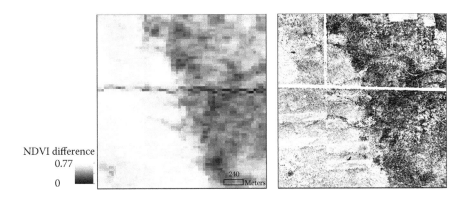

FIGURE 4.3
Seasonal NDVI contrast in a mixed deciduous and conifer forest area at the University of Michigan Biological Station, USA. Left panel shows the difference between leaf-on (summer) and leaf-off (spring) NDVI based on 1998 Landsat 5 TM data (30 m spatial resolution), while right panel shows the same area in an aerial photo (1 m spatial resolution) taken at leaf-off stage in April 1998 with darker areas corresponding to coniferous vegetation (data obtained from USGS Earth Explorer, https://earthexplorer.usgs.gov/).

mathematically finding an inflection point on the curve sides (Melaas et al., 2016) or by selecting consistent cutoff values applicable to a local region, for example, based on a sample of representative pixel level phenologies (White and Nemani, 2006; White et al., 2005).

4.3.3 Hydro-Phenological Analyses of Complex Flooded Landscapes

Single-year phenology is especially critical for differentiating land surface types and dynamic regimes in seasonally variable landscapes, such as wetlands. Changes in water levels and/or vegetation cover may produce mixtures of contrasting cover types at fine spatial scales and thus confusions in single-date mapping approaches. Phenological information from multiple dates may highlight unique flooding schedules, plant communities, and topographic features among wetland cover types that otherwise may not be well known or difficult to assess in the field. Capturing these differences with quantitative phenological indices is also useful for modeling wetland ecosystem function, such as productivity and greenhouse gas sequestration, as shown by a recent study in different-aged restored marshes in California, USA (Knox et al., 2017).

The utility of phenological information for characterizing complex flooded ecosystems was demonstrated by several studies from Poyang Lake, the largest freshwater wetland of China and an important biodiversity hotspot under the international Ramsar convention. Its large size (>3600 km^2), heterogeneous surface, and difficult field access pose major challenges to monitoring of Poyang Lake's habitats and ecosystems with traditional field methods or remote sensing classifications. Multi-temporal remote sensing images thus provide critical hydro-phenological information to differentiate prevalent vegetation types and spatially variable regimes of surface dynamics. Wang et al. (2012) classified dominant plant functional types of Poyang Lake based on their hydro-phenological trajectories from one annual flood cycle and proposed a new time series vegetation-water index based on the seasonal series of NDVI and radar backscatter images. Using similar time series data, Dronova et al. (2015) proposed the framework of "dynamic cover types" as zones of Poyang Lake's prevalent hydro-phenological regimes with characteristic sequences of transitions between flooded, exposed, and/or vegetated states. In Figure 4.4, an example based on the latter study shows how on any given date, at least two and often more dynamic classes may be difficult to separate. Yet, when their full seasonal trajectories are considered, class differences more become apparent, highlighting the importance of multi-temporal phenological information for understanding such complex and dynamic ecosystems.

One important caveat with multi-date phenological analyses is the need to process large numbers of image layers, particularly when multiple time series datasets are used (Neeti and Eastman, 2014). To address this issue, previous studies proposed alternative forms of a classical principal component analysis (PCA) to reduce dimensionality and highlight meaningful changes (Dronova

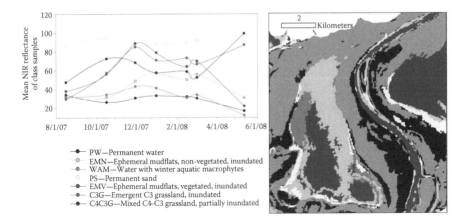

FIGURE 4.4
(**See color insert.**) Example of seasonal trajectories of near infrared reflectance for a set of wetland dynamic cover types at Poyang Lake, China, (left) and their spatial representation in a portion of this wetland area (right).

et al., 2015; Neeti and Eastman, 2014). These include, for instance, a spatial form of PCA orientation which identifies prevalent temporal trajectories from combined multi-temporal pixels or extended PCA (ePCA) which distinguishes prevalent spatio-temporal patterns from combined series of different data sources (Machado-Machado et al., 2011; Neeti and Eastman, 2014). Such PCA-based methods can also reveal previously unknown or unexpected patterns and dynamics and thus provide new information for the future research (Dronova et al., 2015).

4.4 Multi-Year Phenological Inference

4.4.1 Local-Scale: Phenocam Observations

Multi-year phenological analyses need consistent and continuous data for robust inference of seasonal changes and trends. Recent studies have been increasingly utilizing high temporal frequency, fixed view observations of phenology by *in situ* digital cameras, aka "phenocams" (Hufkens et al., 2012; Klosterman et al., 2014; Richardson et al., 2009; Sonnentag et al., 2012; Toomey et al., 2015). Phenocam images can be acquired with various camera types, and are typically collected in visible electromagnetic bands (i.e., red, green and blue) ~every 30 minutes during the active daylight periods (Richardson et al., 2009; Sonnentag et al., 2012). Monitoring changes in vegetation greenness from phenocam photos often relies on RGB-based spectral indices such as green chromatic coordinate, green divergence or simple band ratios (Table 4.1).

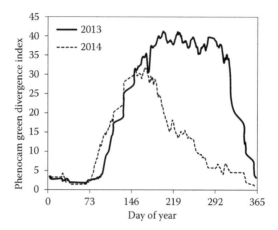

FIGURE 4.5
Seasonal profiles of phenocam greenness index at a freshwater wetland in California, USA, with managed water levels and tall reed (*Typha* & *Schoenoplectus* spp.) vegetation. Lower greenness and earlier senescence of 2014 coincided with plant mortality, browning of leaf tops, and grazing due to cattail caterpillar (*Simyra insularis*) outbreak. Phenocam data courtesy of UC Berkeley's Biometlab.

Phenocam networks are now used all over the world to monitor inter-annual and long term climatic fluctuations. The USA Phenocam network (e.g., https://phenocam.sr.unh.edu/webcam/) provides the data from a variety of participating sites, including biometeorological stations from the AmeriFlux network measuring greenhouse gas fluxes in different ecosystems. Phenocam monitoring has become especially informative and useful in biogeochemical studies (Hufkens et al., 2012; Knox et al., 2017; Toomey et al., 2015) because high frequency temporal variation in vegetation greenness provides important clues about the dynamics of plant cover, photosynthetic activity, disturbances, and other factors affecting ecosystem processes. An example in Figure 4.5 demonstrates the sensitivity of phenocam-based greenness to a sporadic disturbance, where green divergence index of a restored wetland showed lower amplitude and earlier senescence during a pest outbreak in 2014 compared to the previous 2013 season. Strong relationships between phenocam metrics and ecosystem parameters such as gross primary productivity as well as with satellite-based greenness (Hufkens et al., 2012; Knox et al., 2017) support their utility for monitoring ecosystem function in various landscapes as well as validating satellite-based analyses and filling gaps in their time series.

4.4.2 Regional Analyses of Greenness Trends

4.4.2.1 Basic Trend Analyses

One of the most common approaches to detect significant phenological shifts at regional scales is fitting a linear trend to a series of greenness values from similar seasonal stages in different years. For instance, a study in the northern European Fennoscandia region (Hogda et al., 2013) used NDVI-based start of the growing season based on the Advanced Very High Resolution Radiometer (AVHRR) satellite data to detect a warming trend toward earlier season start dates. Such analyses may be also useful for understanding landscape dynamics beyond climatic factors alone; for example, the study of ecosystem recovery following the eruption of Mount St. Helens volcano in Washington, USA, used Landsat-modeled vegetation cover to detect different typologies of recovery dynamics (Lawrence and Ripple, 1999).

A notable example of a continental scale application of time series phenology is represented by the National Dynamic Land Cover Dataset (NDLCD) for Australia (Lymburner et al., 2011). In this effort, remote sensing data have been compiled over ~8 years to produce a national land cover map showing the extent and cover of vegetation at 250×250 m^2 scale. In addition to the map, NDLCD provides the information on the 2000–2008 trends in the Enhanced Vegetation Index (EVI) as the proxy of greenness. Such trends indicate how particular locations and land surface types are changing over time and provide early warning signals of major shifts within specific cover types and localities. Specific parameters such as annual EVI minima, maxima, and trend statistics can be further used in ecosystem models and forecasting of future landscape transitions.

Deviations from statistical trends in the multi-year SVI series (represented by, e.g., residuals of annual or seasonal fit) may also provide important information about landscape change and its drivers. For example, the study in Okavango Delta, Botswana, used the residuals in EVI trends from 14-year Landsat satellite imagery to investigate the signals of climatic cycles, disturbance, and management (Neuenschwander and Crews, 2008). In this study, residuals associated with different periodicities of landscape change provided important indications of both general annual patterns in landscape dynamics and specific events such as landscape flooding and levels of burning and human-induced burning of vegetation.

4.4.2.2 Trajectory-Based Landscape Change Analyses

Detecting and interpreting spatially explicit landscape dynamics from multi-year phenological data has one important challenge: identifying trajectories of similar nature, origin, and schedules that occur at certain temporal lags within the study area. For instance, differences in onset of the same process, such as post-disturbance succession or pest infestation (Kennedy et al. 2007, 2010; Liang et al., 2014a,b) may lead to different shapes of spectral-temporal trajectories within a fixed time frame. To recognize such desynchronized behaviors as similar types of change, various temporal segmentation approaches have been proposed, with a particularly well known Landsat-based LandTrendr tool by Kennedy et al. (2007). The major principle behind temporal segmentation is identifying different sections or time "segments" within the trajectories of multi-temporal pixels and then determining which pixel's trajectories encountered such a segment of change and during which period in the time series. A version of this approach has been also developed for MODIS data (MODTrendr) and successfully applied to detect different types and timings of forest disturbance across the USA's Pacific Northwest region (Sulla-Menashe et al., 2014) and land use changes in Inner Mongolia, China (Yin et al., 2014).

One example of a useful application of temporal segmentation can be found in the 13-year study of forest disturbance and recovery in the southern Rocky Mountains, USA, using Landsat imagery (Liang et al., 2016). For each year, one image with the maximum greenness and least cloud cover was selected to construct the image time series stack. Then the time series trajectory of each pixel was decomposed into a sequence of straight line segments, with each segment representing one salient event happening over that duration. Because different segments possess distinct phenological characteristics, such as slope, duration, magnitude, a random forest classifier was applied subsequently to attribute those segments with the associating events, such as beetle mortality, abrupt fire outbreak or regeneration. In addition to annual disturbance maps, this automated classification workflow simultaneously generates spatially explicit estimates of onset year, severity, and duration of disturbance events (Figure 4.6). These outputs provide critical information for many important ecological questions, including the compound effects of various types of disturbance on the seedling reestablishment, interactions

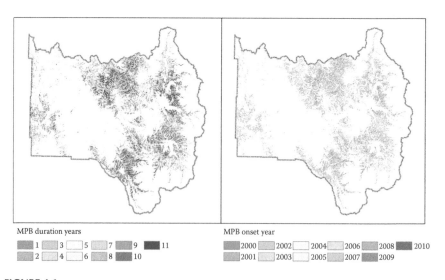

FIGURE 4.6
(**See color insert.**) LandTrendr analysis results showing the duration (left) and onset year (right) of forest mortality attributed to mountain pine beetle infestation. (Based on the study by Liang, L. et al. 2014b. *Remote Sensing* 6, 5696–5716.)

among different disturbances, and driving factors of their agents, such as mountain pine beetle expansion (Liang et al., 2014a).

Similar to the LandTrendr algorithm, another trajectory-based method Vegetation Change Tracker (VCT) also detects changes via simultaneous analysis of all images in a time series but with a different approach (Huang et al., 2010). One highlight of VCT is estimation of a forest z-score and an integrated forest z-score (IFZ) metrics to measure the likelihood of a pixel being a forest pixel and thus identify forested vegetation within a landscape. Additionally, pixels flagged as cloud or cloud shadows are interpolated using clear observations from the neighboring years within a temporal sequence of images. The analysis of forest cover and its change is then based on the physical interpretation of the IFZ temporal profiles over time. This method allows detecting signals of different disturbances and outputs disturbance year and several measures of magnitude to characterize these events.

4.4.2.3 Continuous Change Detection and Classification of Land Cover

Finally, phenological information from detailed image time series may be integrated to detect emerging landscape changes continuously with ongoing collections of satellite data. A recently developed Continuous Change Detection and Classification (CCDC) algorithm based on Landsat data (Zhu and Woodcock, 2014) uses a sinusoid-shaped time series model that distinguishes among land surface dynamics, intra-annual change caused by vegetation phenology, gradual inter-annual change caused by climate variability, vegetation growth or land management, and abrupt changes from noticeable events including deforestation, floods or urbanization. The model is updated when new images are collected, and its parameters are further optimized by applying ordinary least squares method to all the relevant Landsat bands. Within this framework, a land surface change is "confirmed" if the difference between observed and predicted values for a particular pixel exceeds a threshold three consecutive times. In addition to inferring changes, an important advantage of the CCDC method is the possibility to develop and update the land cover classification, because each particular type of land cover typically has a distinct feature set of the time series model and characteristic change behavior.

4.4.3 Broad-Scale Phenological Analyses with Multi-Year Seasonal Data

4.4.3.1 Multi-Year Inference with Phenological Curves

Finally, a critical area of remote sensing-based phenology applications involves detecting continental and global scale transitions in greenness parameters which may signal the effects of climate change and other major factors. Typically, such analyses utilize coarser resolution datasets with global coverage and relatively high temporal frequency, such as AVHRR, MODIS, and Landsat, while specific methods of their processing vary by application.

One popular strategy involves fitting statistical curve functions (such as spline, sigmoid or piecewise linear regression) to evaluate phenological metrics described in Section 4.2 for different years and then investigating whether these parameters show a particular directional shift or anomalous behavior over time. This methodology has been widely applied at regional and even global extents (e.g., Elmore et al., 2012; Melaas et al., 2013; Zhang et al., 2003; Zhang et al., 2006). Some of its applications have developed into a new Landsat-specific phenology algorithm (LPA; Melaas et al., 2013, 2016; Nijland et al., 2014) which estimates the timing of leaf flushing and senescence in deciduous vegetation from the combined Landsat image time series. Because in this framework phenological evaluations are performed separately for each pixel, such analyses also reveal important information about year-to-year phenological variability across the landscape which may indicate seasonal variation in ecosystem functioning and productivity in heterogeneous areas (e.g., Delbart et al., 2005; Polgar and Primack, 2011).

4.4.3.2 Percent above Threshold Approaches

An alternative strategy for multitemporal phenological analyses focuses on phenoregions or climatically similar landscape units, rather than individual pixels, to assess the inter-annual variation in statistical distributions of their greenness (White and Nemani, 2006; White et al., 1997, 2005). In contrast to earlier described methods, this framework advises against curve fitting and instead tracks how many pixels within a region exhibit greenness values above a fixed threshold representing the onset of greening (White and Nemani, 2006). Such a threshold can be statistically evaluated from the times series of a designated index, for example, NDVI for a specific region, and pixels with greenness values above the threshold are summarized as a "percent above threshold" (PAT). Over multiple years, PAT distributions compiled from individual image dates can be used to determine the range of "typical" phenological variation for a given location and to identify anomaly years (White and Nemani, 2006; White et al., 2005). As such, the PAT method avoids the uncertainties associated with statistical curve-fitting; at the same time, its performance may strongly depend on the consistency of long term image availability and thus may be more reliable with high frequency datasets such as AVHRR and MODIS (White and Nemani, 2006).

4.5 Applications and the Importance of Ancillary Factors

Rich multitemporal data archives make it possible to not only assess long term phenological trends but also investigate more specific phenomena and their underlying drivers (Elmore et al., 2012; Friedl et al., 2014). Climatic anomalies are of particular interest because they may trigger unusual responses in vegetation and ecological processes and in some cases may provide early signals of longer term changes in ecosystem function. Both curve-fitting and PAT approaches have been used to analyze such anomalies which may be manifested in outlying deviations of greenness indicators from the "typical" seasonal curves or PAT distributions. For instance, using MODIS time series, Friedl et al. (2014) demonstrated sensitivity of temperate deciduous forest phenology in the northeastern United States to the anomalously warm spring conditions of two recent years. Similarly, White and Nemani (2006) showed how the seasonal trajectory of PAT from a sample phenoregion in Eastern Canada exhibited an earlier-than-typical decline in one of the study years, indicating a late growing season anomaly.

A number of studies have also used phenological time series data to detect the effects of multiple drivers on greenness dynamics, including nonclimatic factors. For example, pronounced long-term declines in spectral greenness across the Mongolian Steppe, one of the world's largest grasslands, appeared to be more strongly connected with grazing-related factors than climatic variation per se based on the analysis of MODIS NDVI time series (Hilker et al., 2014). Similarly, Elmore et al. (2012) used Landsat imagery to investigate the effects of different climate related landscape factors on the timing of spring and autumn vegetation phenology across the mid-Atlantic eastern USA and reported the particular importance of elevation, urban land cover, and distance to tidal water. A study in semi-arid savanna in Southern Africa (Cho et al., 2017) reported the important effects of tree cover on land surface phenological responses from MODIS products which further highlighted spatial variability in ecological characteristics and disturbance risk factors within the savanna ecosystem.

Collectively, these various studies demonstrate an important caveat: apparent correlations among remote sensing-based phenological parameters and climate do not automatically imply a causal relationship, even when climate may be contributing to such variation. Interpreting phenological trajectories and inter-annual shifts requires a rigorous consideration of other potential drivers associated with land use and disturbance as well as the underlying functional processes of plants. A vivid example comes from a recent study in Amazonia (Wu et al., 2016) demonstrating that functional changes in vegetation, namely, new leaf growth patterns, have been the primary cause of photosynthetic seasonality manifested in ecological and remotely sensed patterns, rather than earlier assumed seasonality of climate drivers alone. Thus, applications of plant greenness metrics as phenological proxies need to carefully consider the leaf-level physiological cycles and vegetation configuration within the landscape, particularly at the scale of remote sensing data, rather than simply focusing on pixel-based greenness trends.

Finally, it should be noted that phenological time series analyses become even more in-depth and spatially comprehensive with modern advances in computing platforms and machine learning statistical algorithms. For instance, Clinton et al. (2014) demonstrated important global scale associations of vegetation phenology with climate factors using time series of MODIS EVI, MODIS land surface temperature, and gridded precipitation data using a Google-based coding environment for processing remote sensing data. A web-based Google Earth Engine cloud computing platform provides access to archives of publicly available satellite imagery and geospatial data and facilitates various kinds of phenological analyses from simple greenness index computation to more sophisticated phenology-based classifications and assessments of vegetation and land cover dynamics (e.g., Dong et al., 2016; Simonetti et al., 2015; Wang et al., 2017), also in relation to ecosystem functioning (Knox et al., 2017). As these opportunities continue developing, more informative and robust mining and fusion of extensive remote sensing datasets will enable more accurate diagnoses of phenological phenomena and enable more rigorous applications of phenology in monitoring of climate, land cover-land use and ecosystem services worldwide.

References

Bergen, K.M. and Dronova, I. 2007. Observing succession on aspen-dominated landscapes using a remote sensing-ecosystem approach. *Landscape Ecology* 22, 1395–1410.

Cho, M., Ramoelo, A., and Dziba, L. 2017. Response of land surface phenology to variation in tree cover during green-up and senescence periods in the semi-arid savanna of Southern Africa. *Remote Sensing* 9(7), 689.

Cleland, E.E., Chuine, I., Menzel, A., Mooney, H.A., and Schwartz, M.D. 2007. Shifting plant phenology in response to global change. *Trends in Ecology & Evolution* 22, 357–365.

Clinton, N., Yu, L., Fu, H., He, C., and Gong, P. 2014. Global-scale associations of vegetation phenology with rainfall and temperature at a high spatio-temporal resolution. *Remote Sensing* 6, 7320–7338.

Delbart, N., Kergoat, L., Le Toan, T., Lhermitte, J., and Picard, G. 2005. Determination of phenological dates in boreal regions using normalized difference water index. *Remote Sensing of Environment* 97, 26–38.

Dong, J., Xiao, X., Kou, W., Qin, Y., Zhang, G., Li, L., Jin, C., Zhou, Y., Wang, J., Biradar, C. et al. 2015. Tracking the dynamics of paddy rice planting area in 1986–2010 through time series Landsat images and phenology-based algorithms. *Remote Sensing of Environment* 160, 99–113.

Dong, J., Xiao, X., Menarguez, M.A., Zhang, G., Qin, Y., Thau, D., Biradar, C., and Moore, B. 2016. Mapping paddy rice planting area in northeastern Asia with Landsat 8 images, phenology-based algorithm and Google Earth Engine. *Remote Sensing of Environment* 185, 142–154.

Dronova, I., Bergen, K.M., and Ellsworth, D.S. 2011. Forest canopy properties and variation in aboveground net primary production over upper great lakes landscapes. *Ecosystems* 14, 865–879.

Dronova, I., Gong, P., Wang, L., and Zhong, L. 2015. Mapping dynamic cover types in a large seasonally flooded wetland using extended principal component analysis and object-based classification. *Remote Sensing of Environment* 158, 193–206.

Dymond, C.C., Mladenoff, D.J., and Radeloff, V.C. 2002. Phenological differences in Tasseled Cap indices improve deciduous forest classification. *Remote Sensing of Environment* 80, 460–472.

Elmore, A.J., Guinn, S.M., Minsley, B.J., and Richardson, A.D. 2012. Landscape controls on the timing of spring, autumn, and growing season length in mid-Atlantic forests. *Global Change Biology* 18, 656–674.

Friedl, M.A., Gray, J.M., Melaas, E.K., Richardson, A.D., Hufkens, K., Keenan, T.F., Bailey, A., and O'Keefe, J. 2014. A tale of two springs: using recent climate anomalies to characterize the sensitivity of temperate forest phenology to climate change. *Environmental Research Letters* 9, 54006.

Gitelson, A.A. and Merzlyak, M.N. 1997. Remote estimation of chlorophyll content in higher plant leaves. *International Journal of Remote Sensing* 18, 2691–2697.

Hilker, T., Natsagdorj, E., Waring, R.H., Lyapustin, A., and Wang, Y. 2014. Satellite observed widespread decline in Mongolian grasslands largely due to overgrazing. *Global Change Biology* 20, 418–428.

Hogda, K.A., Tommervik, H., and Karlsen, S.R. 2013. Trends in the start of the growing season in Fennoscandia 1982–2011. *Remote Sensing* 5, 4304–4318.

Huang, C., Coward, S.N., Masek, J.G., Thomas, N., Zhu, Z., and Vogelmann, J.E. 2010. An automated approach for reconstructing recent forest disturbance history using dense Landsat time series stacks. *Remote Sensing of Environment* 114, 183–198.

Huete, A., Didan, K., Miura, T., Rodriguez, E.P., Gao, X., and Ferreira, L.G. 2002. Overview of the radiometric and biophysical performance of the MODIS vegetation indices. *Remote Sensing of Environment* 83, 195–213.

Hufkens, K., Friedl, M., Sonnentag, O., Braswell, B.H., Milliman, T., and Richardson, A.D. 2012. Linking near-surface and satellite remote sensing measurements of deciduous broadleaf forest phenology. *Remote Sensing of Environment* 117, 307–321.

Kennedy, R.E., Cohen, W.B., and Schroeder, T.A. 2007. Trajectory-based change detection for automated characterization of forest disturbance dynamics. *Remote Sensing of Environment* 110, 370–386.

Kennedy, R.E., Yang, Z., and Cohen, W.B. 2010. Detecting trends in forest disturbance and recovery using yearly Landsat time series: 1. LandTrendr - Temporal segmentation algorithms. *Remote Sensing of Environment* 114, 2897–2910.

Klosterman, S.T., Hufkens, K., Gray, J.M., Melaas, E., Sonnentag, O., Lavine, I., Mitchell, L., Norman, R., Friedl, M.A., and Richardson, A.D. 2014. Evaluating remote sensing of deciduous forest phenology at multiple spatial scales using PhenoCam imagery. *Biogeosciences* 11, 4305–4320.

Knox, S.H., Dronova, I., Sturtevant, C., Oikawa, P., Matthes, J., Verfaille, J., and Baldocchi, D.D. 2017. Using digital camera and Landsat imagery with eddy covariance data to model gross primary production in restored wetlands. *Agricultural and Forest Meteorology* 237, 233–245.

Kriegler, F.J., Malila, W.A., Nalepka, R.F., and Richardson, W. 1969. Preprocessing transformations and their effects on multispectral recognition. *Proceedings of the 6th International Symposium on Remote Sensing of Environment*, October 13–16, 1969, Ann Arbor, Michigan, 97–131.

Lawrence, R.L. and Ripple, W.J. 1999. Calculating change curves for multitemporal satellite imagery: Mount St. Helens 1980–1995. *Remote Sensing of Environment* 67, 309–319.

Liang, L., Hawbaker, T.J., Chen, Y., Zhu, Z., and Gong, P. 2014a. Characterizing recent and projecting future potential patterns of mountain pine beetle outbreaks in the Southern Rocky Mountains. *Applied Geography* 55, 165–175.

Liang, L., Chen, Y., Hawbaker, T.J., Zhu, Z., and Gong, P. 2014b. Mapping Mountain pine beetle mortality through growth trend analysis of time-series landsat data. *Remote Sensing* 6, 5696–5716.

Liang, L., Hawbaker, T.J., Zhu, Z., Li, X., and Gong, P. 2016. Forest disturbance interactions and successional pathways in the Southern Rocky Mountains. *Forest Ecology and Management* 375, 35–45.

Lymburner, L., Tan, P., Mueller, N., Thackway, R., Thankappan, M., Islam, A., Lewis, A., Randall, L., and Senarath, U. 2011. *The National Dynamic Land Cover Dataset - Technical report*. Record 2011/031. Geoscience Australia, Canberra. http://www. ga.gov.au/metadata-gateway/metadata/record/71069/

Machado-Machado, E.A., Neeti, N., Eastman, J.R., and Chen, H. 2011. Implications of space-time orientation for Principal Components Analysis of Earth observation image time series. *Earth Science Informatics* 4, 117–124.

McFeeters, S.K. 1996. The use of the normalized difference water index (NDWI) in the delineation of open water features. *International Journal of Remote Sensing* 17, 1425–1432.

Melaas, E.K., Friedl, M.A., and Zhu, Z. 2013. Detecting interannual variation in deciduous broadleaf forest phenology using Landsat TM/ETM plus data. *Remote Sensing of Environment* 132, 176–185.

Melaas, E.K., Sulla-Menashe, D., Gray, J.M., Black, T.A., Morin, T.H., Richardson, A.D., and Friedl, M.A. 2016. Multisite analysis of land surface phenology in North American temperate and boreal deciduous forests from Landsat. *Remote Sensing of Environment* 186, 452–464.

Mickelson, J.G., Civco, D.L., and Silander, J.A. 1998. Delineating forest canopy species in the northeastern United States using multi-temporal TM imagery. *Photogrammetric Engineering and Remote Sensing* 64, 891–904.

Neeti, N. and Eastman, J.R. 2014. Novel approaches in Extended Principal Component Analysis to compare spatio-temporal patterns among multiple image time series. *Remote Sensing of Environment* 148, 84–96.

Neuenschwander, A.L. and Crews, K.A. 2008. Disturbance, management, and landscape dynamics: Harmonic regression of vegetation indices in the lower Okavango Delta, Botswana. *Photogrammetric Engineering and Remote Sensing* 74, 753–764.

Nijland, W., de Jong, R., de Jong, S.M., Wulder, M.A., Bater, C.W., and Coops, N.C. 2014. Monitoring plant condition and phenology using infrared sensitive consumer grade digital cameras. *Agricultural and Forest Meteorology* 184, 98–106.

Ozdogan, M. 2010. The spatial distribution of crop types from MODIS data: Temporal unmixing using Independent Component Analysis. *Remote Sensing of Environment* 114, 1190–1204.

Polgar, C.A. and Primack, R.B. 2011. Leaf-out phenology of temperate woody plants: From trees to ecosystems. *New Phytologist* 191, 926–941.

Reed, B., Brown, J., Vanderzee, D., Loveland, T., Merchant, J., and Ohlen, D. 1994. Measuring phenological variability from satellite imagery. *Journal of Vegetation Science* 5, 703–714.

Richardson, A.D., Anderson, R.S., Arain, M.A., Barr, A.G., Bohrer, G., Chen, G., Chen, J.M., Ciais, P., Davis, K.J., Desai, A.R. et al. 2012. Terrestrial biosphere models need better representation of vegetation phenology: results from the North American Carbon Program Site Synthesis. *Global Change Biology* 18, 566–584.

Richardson, A.D., Braswell, B.H., Hollinger, D.Y., Jenkins, J.P., and Ollinger, S.V. 2009. Near-surface remote sensing of spatial and temporal variation in canopy phenology. *Ecological Applications* 19, 1417–1428.

Rouse, J., Haas, R., Scheel, J., and Deering, D. 1974. Monitoring Vegetation Systems in the Great Plains with ERTS. *Proceedings, 3rd Earth Resource Technology Satellite (ERTS) Symposium*, December 10–14, 1973, Washington, DC, 1, 48–62.

Sakamoto, T., Yokozawa, M., Toritani, H., Shibayama, M., Ishitsuka, N., and Ohno, H. 2005. A crop phenology detection method using time-series MODIS data. *Remote Sensing of Environment* 96, 366–374.

Simonetti, D., Simonetti, E., Szantoi, Z., Lupi, A., and Eva, H.D. 2015. First Results from the Phenology-Based Synthesis Classifier Using Landsat 8 Imagery. *Ieee Geoscience and Remote Sensing Letters* 12, 1496–1500.

Sonnentag, O., Detto, M., Vargas, R., Ryu, Y., Runkle, B.R.K., Kelly, M., and Baldocchi, D.D. 2011. Tracking the structural and functional development of a perennial pepperweed (Lepidium latifolium L.) infestation using a multi-year archive of webcam imagery and eddy covariance measurements. *Agricultural and Forest Meteorology* 151, 916–926.

Sonnentag, O., Hufkens, K., Teshera-Sterne, C., Young, A.M., Friedl, M., Braswell, B.H., Milliman, T., O'Keefe, J., and Richardson, A.D. 2012. Digital repeat photography for phenological research in forest ecosystems. *Agricultural and Forest Meteorology* 152, 159–177.

Sulla-Menashe, D., Kennedy, R.E., Yang, Z., Braaten, J., Krankina, O.N., and Friedl, M.A. 2014. Detecting forest disturbance in the Pacific Northwest from MODIS time series using temporal segmentation. *Remote Sensing of Environment* 151, 114–123.

Toomey, M., Friedl, M.A., Frolking, S., Hufkens, K., Klosterman, S., Sonnentag, O., Baldocchi, D.D., Bernacchi, C.J., Biraud, S.C., Bohrer, G. et al. 2015. Greenness indices from digital cameras predict the timing and seasonal dynamics of canopy-scale photosynthesis. *Ecological Applications* 25, 99–115.

Wang, J., Xiao, X., Qin, Y., Dong, J., Geissler, G., Zhang, G., Cejda, N., Alikhani, B., and Doughty, R.B. 2017. Mapping the dynamics of eastern redcedar encroachment into grasslands during 1984-2010 through PALSAR and time series Landsat images. *Remote Sensing of Environment* 190, 233–246.

Wang, L., Dronova, I., Gong, P., Yang, W., Li, Y., and Liu, Q. 2012. A new time series vegetation-water index of phenological-hydrological trait across species and functional types for Poyang Lake wetland ecosystem. *Remote Sensing of Environment* 125, 49–63.

White, M.A., Hoffman, F., Hargrove, W.W., and Nemani, R.R. 2005. A global framework for monitoring phenological responses to climate change. *Geophysical Research Letters* 32, L04705.

White, M.A. and Nemani, R.R. 2006. Real-time monitoring and short-term forecasting of land surface phenology. *Remote Sensing of Environment* 104, 43–49.

White, M.A., Thornton, P.E., and Running, S.W. 1997. A continental phenology model for monitoring vegetation responses to interannual climatic variability. *Global Biogeochemical Cycles* 11, 217–234.

Woebbecke, D., Meyer, G., Vonbargen, K., and Mortensen, D. 1995. Color indexes for weed identification under various soil, residue, and lighting conditions. *Transactions of the Asae* 38, 259–269.

Wolter, P., Mladenoff, D., Host, G., and Crow, T. 1995. Improved forest classification in the northern Lake-states using multitemporal Landsat imagery. *Photogrammetric Engineering and Remote Sensing* 61, 1129–1143.

Wu, J., Albert, L.P., Lopes, A.P., Restrepo-Coupe, N., Hayek, M., Wiedemann, K.T., Guan, K., Stark, S.C., Christoffersen, B., Prohaska, N. et al. 2016. Leaf development and demography explain photosynthetic seasonality in Amazon evergreen forests. *Science* 351, 972–976.

Xiao, X.M., Hollinger, D., Aber, J., Goltz, M., Davidson, E.A., Zhang, Q.Y., and Moore, B. 2004. Satellite-based modeling of gross primary production in an evergreen needleleaf forest. *Remote Sensing of Environment* 89, 519–534.

Yin, H., Pflugmacher, D., Kennedy, R.E., Sulla-Menashe, D., and Hostert, P. 2014. Mapping Annual Land Use and Land Cover Changes Using MODIS Time Series. *Ieee Journal of Selected Topics in Applied Earth Observations and Remote Sensing* 7, 3421–3427.

Zhang, X., Friedl, M.A., and Schaaf, C.B. 2006. Global vegetation phenology from Moderate Resolution Imaging Spectroradiometer (MODIS): Evaluation of global patterns and comparison with *in situ* measurements. *Journal of Geophysical Research-Biogeosciences* 111, G04017.

Zhang, X.Y., Friedl, M.A., Schaaf, C.B., Strahler, A.H., Hodges, J.C.F., Gao, F., Reed, B.C., and Huete, A. 2003. Monitoring vegetation phenology using MODIS. *Remote Sensing of Environment* 84, 471–475.

Zhong, L., Hawkins, T., Biging, G., and Gong, P. 2011. A phenology-based approach to map crop types in the San Joaquin Valley, California. *International Journal of Remote Sensing* 32, 7777–7804.

Zhong, L., Gong, P., and Biging, G.S. 2012. Phenology-based crop classification algorithm and its implications on agricultural water use assessments in California's central valley. *Photogrammetric Engineering and Remote Sensing* 78, 799–813.

Zhu, Z. and Woodcock, C.E. 2014. Continuous change detection and classification of land cover using all available Landsat data. *Remote Sensing of Environment* 144, 152–171.

Zipper, S.C., Schatz, J., Singh, A., Kucharik, C.J., Townsend, P.A., and Loheide, S.P. 2016. Urban heat island impacts on plant phenology: Intra-urban variability and response to land cover. *Environmental Research Letters* 11, 54023.

5

Time Series Analysis of Moderate Resolution Land Surface Temperatures

Benjamin Bechtel and Panagiotis Sismanidis

CONTENTS

5.1 Introduction 89
5.2 Data and Methods 92
    5.2.1 MYD11A1 and MOD11A1 Land Surface Temperatures 92
    5.2.2 MODIS Land Cover and Urban Areas 94
    5.2.3 Annual Cycle Parameters 94
5.3 Results and Discussion 96
    5.3.1 ACP for Central Europe 96
    5.3.2 Comparison between Collection-5 and Collection-6 98
    5.3.3 Latitudinal Gradients in ACP 102
5.4 Applications 107
    5.4.1 Climatological SUHI Analysis 107
    5.4.2 Using the ACPs as Disaggregation Kernels
         for Downscaling LST Image Data 112
5.5 Conclusions 115
References 115

5.1 Introduction

The land surface temperature (LST) is a key parameter in the physics of the Earth that drives the radiative, latent, and sensible heat fluxes at the surface-atmosphere interface (Guillevic et al., 2012; Li et al., 2013a; Townshend et al., 1994). It is a highly dynamic parameter that changes rapidly through space and time due to variations in the Earth's surface, weather, and also due to variations in the incoming solar radiation. Accurate knowledge of LST can provide information about the surface equilibrium state, which is of paramount importance for a number of fields including hydrology, climatology, and geophysics (Bechtel, 2015a; Li et al., 2013a). LST data can also be used for the study of urbanization and its impact on the urban thermal environment, namely the surface urban heat island (SUHI) (Keramitsoglou et al., 2011, 2012; Voogt and Oke, 2003; Weng, 2009).

Presently, the only practical way to acquire spatially detailed, global, accurate (1–2 °C), and frequent LST measurements is with satellite remote sensing of thermal infrared radiation (TIR) (Li et al., 2013a). Most satellite sensors acquire TIR data over the 8 to 13 μm atmospheric window that depends on the LST according to Planck's equation. In particular, under clear sky conditions, the TIR signal measured by a satellite sensor consists of three major radiation components (Schowengerdt, 2006). These components, which originate both from the Earth's surface and atmosphere, are: (1) the surface-emitted radiation; (2) the atmospheric TIR radiation that is reflected by the Earth's surface in the instantaneous field-of-view (IFOV) of the sensor; and (3) the path-emitted TIR radiation. To retrieve the LST from TIR satellite data requires the surface emitted radiation to be decoupled from the two atmospheric components (components 2 and 3) and adjusted for the atmospheric attenuation and the emissivity of the Earth's surface. However, the land surface emissivity (LSE) is usually unknown, thus the estimation of the LST from satellite measurements is mathematically unsolvable (for N radiance measurements the unknowns are $N + 1$, i.e., the emissivity of each spectral band plus the LST) (Li et al., 2013a). To make the LST retrieval possible, the LSE has to be known *a priori* or the LST and LSE have to be simultaneously solved using various constraints and assumptions.

Over the past decades, several satellite LST retrieval algorithms that require *a priori* knowledge of LSE have been developed (Li et al., 2013a). These algorithms can be divided into three main categories: (i) the single-channel algorithms (Jimenez-Munoz and Sobrino, 2010; Qin et al., 2001; Sobrino et al., 2004), (ii) the multi-channel algorithms (Price, 1984; Sobrino et al., 1996; Wan and Dozier, 1996), and (iii) the multi-angle algorithms (Sobrino et al., 1996). The algorithms of the first category estimate the LST by solving the radiative transfer equation (RTE). To do so, they require accurate knowledge of the atmospheric conditions so as to correct the surface emitted TIR radiance for the atmospheric effects. However, this information is hard to get. The algorithms of the second category, also known as split-window (SW) algorithms, require the use of at least two adjacent spectral bands and estimate the LST as a combination of the brightness temperatures of these TIR bands (Li et al., 2013a; Wan and Dozier, 1996). In contrast to single-channel algorithms, SW algorithms do not require information about the atmospheric conditions, but empirically correct the atmospheric effects using the differential atmospheric absorption of the adjacent TIR spectral bands. The multi-angle algorithms share the same concept as the multi-channel algorithms but use the differential atmospheric absorption of different viewing paths instead of adjacent spectral bands (Li et al., 2013a).

The LSE information required can be estimated *a priori* using various LSE retrieval methods (Li et al., 2013b). Among the most widely used, are (i) the classification based methods (Mitraka et al., 2012; Snyder et al., 1998), which assume a constant emissivity within a particular land cover class; and (ii) the normalized difference vegetation index (NDVI) based methods (Sobrino et al., 2008; Valor and Caselles, 1996), which assume the vegetation fraction to be the dominant influence on the emissivity. Alternatively, LSE and LST can be simultaneously retrieved. Presumably the most relevant method for simultaneous LST and LSE retrieval is the temperature emissivity separation (TES) algorithm proposed in Gillespie et al. (1998) for ASTER (Advanced Spaceborne Thermal Emission and Reflection Radiometer), which requires three or more TIR bands. Overall, the availability of high quality LSE data is crucial for the accurate retrieval of LST, since a 1% uncertainty in LSE can lead to a 0.5 K error in LST (Li et al., 2013a).

Even after atmospheric and emissivity correction, remote sensing only allows estimation of the directional radiometric surface temperature or skin temperature, which is an approximation of the thermodynamic temperature (*in situ* temperature of the solid mass at the interface between surface and atmosphere) (Becker and Li, 1995; Norman and Becker, 1995). Moreover, the interpretation of such data is very difficult. This is because the definition of surface temperature over heterogeneous nonisothermal surfaces is ambiguous. This is an important problem in TIR remote sensing and is more pronounced in coarse resolution sensors due to the mixed pixel effect. Furthermore, the use of remotely sensed LST data is further hampered because currently available remote sensors cannot acquire data that capture the diurnal evolution of LST at an appropriate spatial scale (100 m) (Inamdar et al., 2008; Sismanidis et al., 2015; Zakšek and Oštir, 2012). To that end, clouds further reduce the availability of satellite TIR data, especially over large areas in humid and moderate climates (Bechtel, 2015a,b). Hence, new methods for using TIR remote sensing data are required.

A method to overcome these limitations was proposed in Bechtel (2011, 2012, 2015a,b) and involves the monitoring of the annual LST cycle by time series analysis. The annual LST cycle is periodic and primarily depends on the incoming solar irradiation and the inclination of the Earth's axis to the ecliptic. In brief, Bechtel (2011, 2012, 2015a,b) proposed to fit a sine model to a multi-year LST time series for a pixel and then to use the parameters that define the fitted model and retain the geometry of the original LST data. These parameters, which are known as the LST annual cycle parameters (ACPs), are the Mean Annual Surface temperature (MAST), the Yearly Amplitude of Surface Temperature (YAST), and the phase shift from the spring equinox (Theta). Additionally, the root mean squared error (RMSE) of the fit and the number of clear sky LST acquisitions (NCSA) are employed. The ACP method uses satellite LST data in a novel way and enables a continuous description of the thermal landscape which can then be used in numerous applications, such as the study of SUHIs (Bechtel, 2015b; Fu and Weng, 2016; Huang et al., 2016), geothermal energy (Hein et al., 2016), water masking (Klein et al., 2017), ecological modelling (Bobrowski et al., In review), topo-climatology (Bechtel, 2016) climate classification, and LST downscaling (Bechtel et al., 2012; Sismanidis et al., 2016, 2017; Zhan et al., 2016) to name a few. Limitations comprise the insensitivity towards short term fluctuations and extreme conditions, a bias towards cloud free conditions, and the infeasibility of the approach in the tropics.

This chapter is concerned with the ACPs and presents new improvements, further evaluation tests, and potential applications. Section 5.2 presents the ACP calculation in detail. In Section 5.3 the ACPs of Central Europe are presented and discussed, while a land cover analysis of the differences between the ACPs estimated using collection-5 MODIS LST (presented initially in Bechtel, 2015a,b) and collection-6 is presented. Section 5.3 includes an analysis of the latitudinal (70° N to 35° S) gradients so as to assess the distribution and performance of the ACPs for various latitudes. Based on this analysis, new improvements in the ACP methodology are proposed. In Section 5.4 two key applications where ACPs can prove valuable are discussed, namely the study of the SUHI and the increase of the spatial resolution of coarse-scale LST data (e.g., LST from geostationary satellite sensors). Finally, the drawn conclusions are presented in Section 5.5.

5.2 Data and Methods

5.2.1 MYD11A1 and MOD11A1 Land Surface Temperatures

The data used in this work are the collection-5 and collection-6 daily Aqua and Terra MODIS LST time series (i.e., MYD11A1 and MOD11A1, respectively) and the tiles and times for the conducted experiments are presented in Table 5.1. The employed datasets cover Europe and Africa (Figure 5.1; region of interest (ROI) 1 and ROI 2, respectively) at a spatial resolution of 1 km (precisely 0.928 km) and are freely available from the U.S. Geological Survey (USGS) Land Processes Distributed Active Archive Center (LP DAAC), which is a component of NASA's Earth Observing System Data and Information System (EOSDIS). Each data product is delivered in a gridded hierarchical

TABLE 5.1

Conducted Experiments and Data Processed for this Study

Experiment	Region (tiles)	Satellites	Times	Collection	Years	ACP Models
Compare MODIS collections v5 and v6	ROI1A (h17v03-h19v05)	Aqua	day, night	v5 and v6	2003–2014	ACP3
Analyse model performance in low and high latitudes	ROI2 (h19v01-h19v12)	Aqua	day	v6	2011–2015	ACP3, ACP5
Analyse surface urban heat islands	ROI1B (h17v03-h19v04)	Aqua, Terra	day, night	v6	2011–2015	ACP3

Region refers to the regions of interest presented in Figure 5.1.

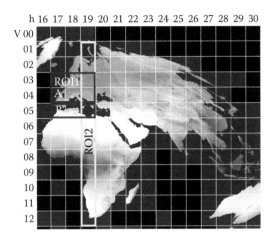

FIGURE 5.1
Tiles and projection of MODIS land products and selected subsets: central Europe (ROI 1A: h17v03—h19v04) with Northern Africa (ROI 1B: h17v03—h19 v05), and latitudinal gradient (ROI 2: h19v01—h19v12).

data format (HDF) file that contains a number of scientific datasets, namely the LST; the 11 μm (MODIS band 31) and 12 μm (MODIS band 32) LSE; a quality control flag; the viewing time (in local solar time); the viewing angle; and the number of cloud free acquisitions that were used for calculating the LST (it might be more than one due to overlapping orbits).

Aqua and Terra satellites are on a sun-synchronous, near-polar, circular orbit with equatorial crossing times in the descending (daytime) orbit around 13:30 hrs and 10:30 hrs (local solar time), respectively, and in the ascending (night time) orbit around 01:30 hrs and 22:30 hrs, respectively. Hence, each HDF file contains two LST datasets, one for daytime and one for night time (this is also the case for the other supplementary datasets, e.g., the quality flag and the viewing time and angle, but not for the LSE, which is common between the daytime and night time acquisition).

In this work, both daytime and night time LSTs are employed. Even though a few days are missing, the employed time series are almost complete (mostly more than 99% of days). Moreover, it has to be noted that the viewing time and viewing angle of the MODIS LST data are not always identical. In particular, the acquisition time might differ by more than 1 hr and the viewing angle can range from −55° to +55°. However, because the number of observations employed is large, these effects have been assumed negligible.

The LST data of MYD11A1 and MOD11A1 are retrieved using information from MODIS bands 31 (10.78–11.28 μm) and 32 (11.77–12.27 μm) through a generalized SW algorithm (Wan and Dozier, 1996) that requires *a priori* knowledge of the LSE. The LSE information for these two bands is retrieved from a classification based approach that considers 18 different land cover types and has been presented in Snyder et al. (1998) and refined in Wan (2008) and

Wan (2014). The accuracy of the LST product has been validated and was found to mostly fall within 1 K (Wan et al., 2004; Wan, 2008; Wang and Liang, 2009; Wan, 2014). However, for urban areas, lower LST accuracy is expected. This is due to increased uncertainties in the emissivity of the urban materials and the geometric effects induced by the three-dimensional structure of the urban surfaces (Krayenhoff and Voogt, 2016; Voogt and Oke, 1998; Yang et al., 2015).

5.2.2 MODIS Land Cover and Urban Areas

The MODIS Land Cover product MCD12Q1 version 5.1 (Friedl et al., 2010) provides annual global land cover classifications in 500 m resolution derived from both Terra and Aqua data. Here, the first classification scheme defined by the International Geosphere Biosphere Programme (IGBP) is used, which comprises 17 land cover classes (11 natural vegetation classes, three developed and mosaicked land classes, and three nonvegetated land classes). In this work, the employed MCD12Q1 land cover data (2013) were used (i) for analyzing the differences between the collection-5 and collection-6 ACPs of ROI 1A + B (Figure 5.1) and (ii) for extracting the SUHIs of ROI 1A (Figure 5.1).

For analyzing the differences between the collection-5 and collection-6 ACPs, the corresponding MCD12Q1 data were resampled (using nearest neighbor) to 1 km and then reclassified to four major classes: forest areas (IGBP classes 1, 2, 3, 4 and 5); croplands (IGBP classes 10, 12 and 14); bare soil (IGBP class 16), and built-up (IGBP class 13). The remaining classes cover a very small fraction of the study area and thus they were not included in the analysis.

For SUHI extraction only every second column and line were read for resolution consistency. Then a mask of the urban class was derived and major urban centers were extracted by a morphological opening with a disk shaped structuring element of two pixels radius, thereby excluding small towns and villages. The corresponding rural areas were defined as all pixels within 15 pixels (~15 km) distance from the urban core areas that are not urban themselves.

5.2.3 Annual Cycle Parameters

The ACPs use long LST time series to generate a pixel-wise simplified climatology (Bechtel, 2015a,b). This allows the use of information contained in scenes with partial cloud which otherwise would be discarded. Especially for high latitude and tropical regions, the number of cloud contaminated scenes is typically large, as Figure 5.2 presents for the case of Brussels in Belgium. In particular, Figure 5.2 displays the 2013 time series of daily MODIS daytime LST over Brussels and shows that even on small scale only a few scenes are entirely cloud free.

To derive the annual cycle, a simple model (referred to henceforth as ACP3) given in Equation 5.1 is fitted to the time series.

LST (K) ⟵ 275   285   295   305 ⟶

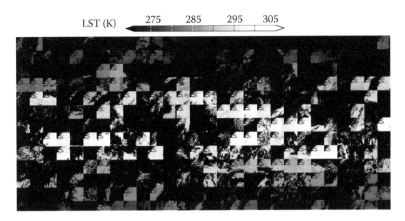

FIGURE 5.2
One year of Terra MODIS daytime LST (in K) observations for Brussels: Jan. 1st (upper left) to Dec. 30th (lower right) 2013.

$$T_s(d) = MAST + YASTsin(k(d + \theta)), \quad \text{with} \quad k = \frac{2\pi}{365} \tag{5.1}$$

In addition to the three model parameters, three additional parameters are computed. The RMSE of the fit is the quadratic mean of the residuals and provides an integrated measure of both the inter-diurnal and inter-annual variability of LST. The R^2 provides the coefficient of determination of the model fit and hence an indicator of the model's accuracy; finally, the number of clear sky acquisitions (NSCA) used is a measure for the frequency of cloud presence (Figure 5.3).

It has been stated in Bechtel (2015a,b) that in the tropics, the ACP model fit is less meaningful, since the annual cycle of irradiation in the tropics is characterized by two annual maxima instead of one. Therefore, in this work we suggest an alternative ACP model (referred henceforth as ACP5) with a second harmonic function, as given in Equation 5.2.

$$T_s(d) = C + A_1 sin(k_1(d + \theta_2)) + A_2 sin(k_2(d + \theta_2)),$$
$$\text{with} \quad k_1 = \frac{2\pi}{365} \quad \text{and} \quad k_2 = \frac{4\pi}{365} \tag{5.2}$$

In this equation, C is the constant part (equivalent to $MAST$), A_1 (equivalent to $YAST$) and A_2 are the amplitudes of the annual and biannual variations, and θ_1 and θ_2 are the respective phase shifts.

The ACP were estimated from the MODIS data presented in Table 5.1 using an unconstrained nonlinear optimization algorithm minimizing the square sum of the residuals using parallel processing on a server with 4 sockets of AMD Opteron (Abu-Dhabi) 6386 SE (16 cores each, 1.4 to 2.8 GHz),

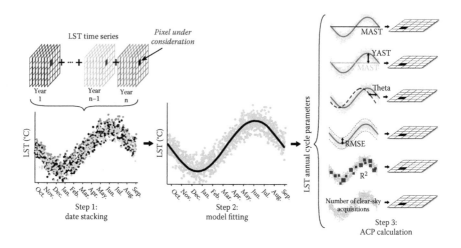

FIGURE 5.3

Principle of the annual cycle parameter (ACP) calculation. Step 1: Sequence of acquisitions from one pixel are stacked into a time series. Step 2: A annual cycle model is fit to the time series. Step 3: The model parameters are extracted for each pixel: MAST, YAST, Theta, RMSE, R^2, and NCSA.

256 GB RAM, and a 100 TB external general parallel file system (for details see Bechtel, 2015a,b). For the preliminary implementation of the ACP5 model, the algorithm was considerably more computing intensive and did not always converge, hence different optimization algorithms will be considered in the future.

5.3 Results and Discussion

5.3.1 ACP for Central Europe

The patterns of the ACP3 represent robust and gap free estimates of surface thermal characteristics and reveal various influencing factors, as presented in Figure 5.4 for Central Europe.

In particular, MAST (Figure 5.4a) shows a strong imprint of orography and a latitudinal gradient (see following section). Additionally, cities are well visible and clearly warmer than their surroundings (see Section 4.3). YAST (Figure 5.4b) shows a strong East-West gradient, indicating the higher seasonality in the more continental climates in the East. In addition, a strong imprint of land cover can be seen, with higher annual amplitudes in the arid areas of Spain. Theta (Figure 5.4c) shows a large delay of mountainous regions with extended snow cover as well as water bodies. The RMSE (Figure 5.4d) is largely influenced by the annual inter-variation which is lower in the maritime climates (e.g., Great Britain) and larger in

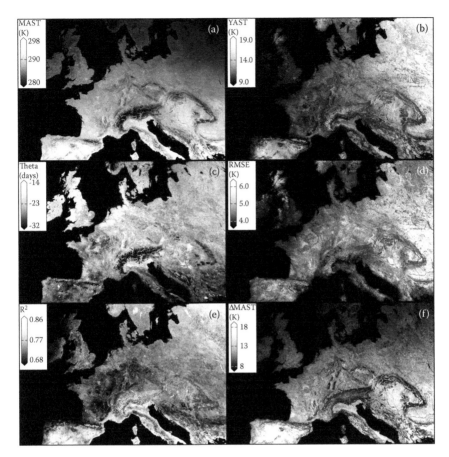

FIGURE 5.4
Annual cycle parameters for Central Europe, Aqua daytime, 2011–2015. (a) MAST, (b) YAST, (c) Theta, (d) RMSE, (e) R^2, (f) ΔMAST: daytime minus night time MAST.

the more continental climates in Eastern Europe. R^2 (Figure 5.4e) shows large accordance with the amplitude, indicating better model performance. However, it is also greatly influenced by land cover, with lakes achieving the best and forest substantially worse results (clearly visible as the dark area that extends over Southern France to Central Germany). The NCSA (not shown in Figure 5.4) mostly reflect the cloud frequency with higher values in the Mediterranean and lower values in Northern Europe. In addition, river pixels have more acquisitions, which is a new characteristic of the MODIS collection 6 LST data (see Section 3.2). ΔMAST (Figure 5.4f) is the difference between daytime and night time MAST and thus a proxy for the diurnal temperature range. Compared to MAST, the effects of land cover and topography are even more visible, since the latitudinal gradient is partly removed by the substraction. Additional features become visible at higher zoom levels.

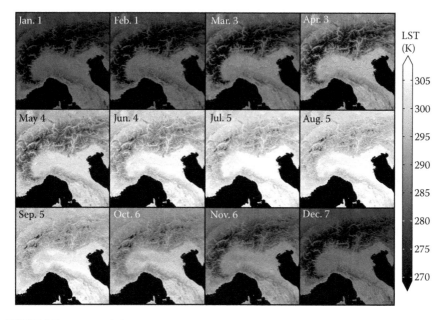

FIGURE 5.5
Seasonal changes of virtual LST pattern for 13:30 generated from the ACP model for Northern Italy and the Alps.

In addition to the ACP themselves, the annual cycle models can be used to generate virtual LST patterns for every day of year. This can be understood as the climatological surface temperature without inter-daily and inter-annual fluctuations. An example for the Alps is given in Figure 5.5. It can be clearly seen that the predicted LST patterns change over the year due to the phase shift of the annual cycle. For instance, the LST difference between the Alps and the Northern surroundings is very pronounced in spring and almost vanishes in September and October when only a few glaciers and areas above the climatic snow line remain dark. Likewise, the included SUHIs are very pronounced in summer but almost invisible in winter (see Section 4.1).

5.3.2 Comparison between Collection-5 and Collection-6

The first version of the ACPs, presented in Bechtel (2015a,b), has been produced using as input the collection-5 (Wan, 2007) daily MODIS LST data products (i.e., the MOD11A1 and MYD11A1). However, the MODIS collection-5 validation activities showed that these LST data products considerably underestimated the LST of desert and bare soil areas (errors ranging from 2 to 4.5 K were observed at the In-salah validation site in Algeria; Wan, 2014). These errors were attributed mostly to LSE as well as the employed SW algorithm, which did not sufficiently cover the wide range of bare soil LST (Wan, 2014). To address this problem, the recently released (2015–2016) collection-6 has implemented two key refinements in the LST production code. These two refinements are: (i) the use of two sets of SW coefficients for retrieving the LST of bare soil areas between latitudes 39° S and 49.5° N (the first one for daytime data and the second for night time); and (ii) the refinement of band 31 and band 32 LSEs by adjusting their difference without changing their mean value (Wan, 2015). Overall, the validation of collection-6 MODIS LST data showed superior results over bare soil and desert areas in respect to collection-5 (Wan, 2015).

Figure 5.6 shows a color composite of the three ACP3 from collection-5 (left) and collection-6 (right). It shows that the general pattern is very consistent between the two versions and the spatial variations are much larger. However, some smaller differences can be found, which are discussed in the following paragraphs.

To assess how the refinements of collection-6 affect the ACP3 data, a comparison between the collection-5 and the collection-6 MYD11A1 daytime MAST, YAST, Theta, RMSE, and NCSA over Europe and North Africa was performed. The results were analyzed using land cover information from MODIS MCD12Q1 data product and are presented in the maps of Figure 5.7 and the boxplots of Figure 5.8. As expected, the greatest differences are found over the Sahara desert in North Africa. The ACPs most affected by the algorithm refinements are MAST, YAST, and NCSA. In particular, a considerable increase of MAST of about +2.5 K is evident over the Sahara desert (Figure 5.7a), while for the forest areas, croplands, and built-up areas over Europe the change between collection-5 and collection-6 MAST is minor (close to +0.4 K; Figure 5.8a). YAST does not exhibit an equally extensive change as MAST and the largest differences (+1.4 K) are mostly confined over the Algerian desert. For the rest of the examined land cover types, the change in YAST is close to +0.4 K. The change in Theta and RMSE are minor and close to 0 as Figure 8c and d reveal.

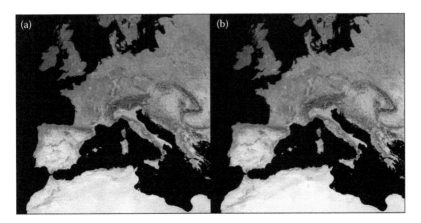

FIGURE 5.6
(**See color insert.**) RGB composite of (a) collection-5 (YAST, MAST, Theta) and (b) collection-6 (YAST, MAST, Theta).

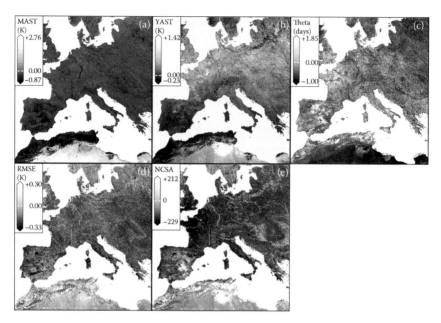

FIGURE 5.7
Maps of the differences between collection-6 and -5 Aqua daytime ACP3.

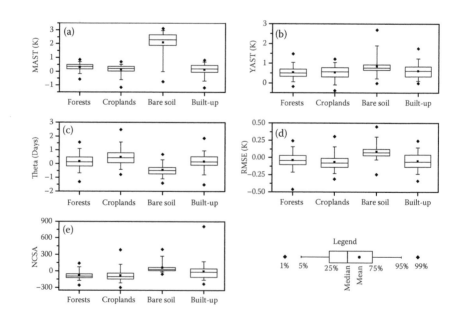

FIGURE 5.8
The distribution of the differences between the collection-6 and collection-5 Aqua daytime ACP3 for selected land covers.

Over the Sahara, Theta slightly decreases by 1 day (Figure 5.7c), while over some areas in Spain it slightly increases by +1.8 days. The RMSE change map (Figure 5.7d) does not exhibit any particular hotspots and the changes between collection-5 and collection-6 are uniformly distributed over the study area. Besides MAST and YAST, the collection-6 algorithm refinements also affect the NCSA parameter. In particular, Figures 7e and 8e reveal that there is a slight decrease of the clear sky pixels over Europe (mostly forest and cropland areas) and an increase over bare soil areas (the mean is +58 acquisitions). For built-up areas, the mean change in NCSA is close to 0, but the distribution (Figure 5.8e) of the collection-5 and collection-6 NCSA differences exhibit a long positive tail (i.e., the 95th and 99th percentile are equal to 173 and 807, respectively), which means that for some pixels the number of available acquisitions increases considerably.

A case of considerable increase in NCSA is the city of Paris in France. To that end, in Figure 5.9a, the corresponding NCSA (2005–2014) parameter is presented both for collection-5 and collection-6 Aqua daytime LST. The comparison indicates that the collection-6 NCSA decreases for the areas surrounding Paris, but for the city of Paris it increases considerably (for some pixels the increase was approximately +1000 acquisitions).

The time series for a selected pixel (Figure 5.9b) shows the reason for this. For the daytime, collection-5 clear-sky acquisitions are available only for April till October and during winter months the few available measurements are outliers. Nevertheless, the ACP3 fit is quite robust and mostly compensates for the

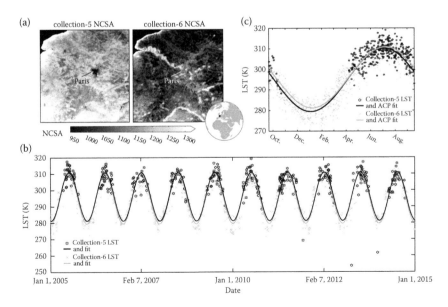

FIGURE 5.9
(a) NCSA for Paris (2005–2014), Aqua daytime, collection-5 and -6. (b) Time series; and (c) the annual cycle for selected pixel.

missing data (the collection-5 and collection-6 fits in Figure 5.9c are remarkably similar). This observation is in line with Bechtel (2012). The above findings suggest that collection-6 MODIS LSTs are better suited for SUHI studies than collection-5 and that the NCSA is generally decreasing between collection-5 and collection-6. For all subsequent analysis, collection-6 data were used.

5.3.3 Latitudinal Gradients in ACP

Figure 5.10 presents the ACP3 for ROI 2 (Figure 5.1), which is a latitudinal gradient from Scandinavia, over Eastern Europe, Saharan Africa (especially Libya), and tropical Africa (Cameroon, Gabon, Congo), toward the savannas

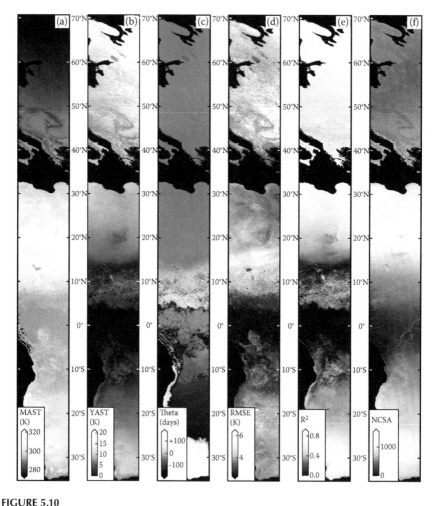

FIGURE 5.10
The ACPs (Aqua daytime, 2011–2015, ACP3) at different latitudes. (a) MAST, (b) YAST, (c) Theta, (d) RMSE, (e) R², and (f) NSCA.

and deserts in Southern Africa (Angola, Namibia, and the Republic of South Africa). The transect is ideally suited to study the performance of ACP3 at different latitudes. All numbers given in the following paragraphs refer to 5° increments of latitude, 10° thus representing the average of 7.5–12–5° N, as shown in Table 5.2.

MAST (Figure 5.10a) shows a strong latitudinal gradient but also substantial impact of land cover. In particular, the arid areas of Sahara (~20°N) and Namib (~25°S) exhibited the highest MAST values (~319.0 K and ~315.0 K, respectively) and are warmer than the rainforest areas in the tropics (approx. between 10°S to 10°N). The lowest MAST values are found at 80 °N (256.6 K) in Northern Europe. Orographic features also influence MAST, as it can be seen in Figure 5.10a for the Carpathian Mountains in Romania (~45°N). In contrast to MAST, YAST (Figure 5.10b) shows an inverted latitudinal gradient with the annual amplitude increasing with latitude. This reflects the stronger seasonality at higher latitudes. To that end, the lowest YAST values (0.4 K) are found at the equator and the highest at 65° N (20.0 K).

Theta (Figure 5.10c), corresponding to the phase shift relative to the spring equinox, does not exhibit a strong latitudinal gradient, but most variations

TABLE 5.2

Latitudinal Changes of ACP3 Values Based on h19v01–12 Tiles in 5° Steps (+ Is North and − Is South)

Latitude (°)	MAST (K)	YAST (K)	Theta (days)	RMSE (K)	R^2	NSCA (days)	NSCA (%)
65.0	275.6	20.0	−26.1	5.4	0.88	603	0.33
60.0	279.2	18.0	−23.9	4.9	0.86	521	0.29
55.0	283.5	16.9	−19.7	5.4	0.80	525	0.29
50.0	289.0	15.4	−20.1	5.5	0.77	645	0.35
45.0	292.8	13.7	−23.9	5.3	0.75	714	0.39
40.0	295.5	13.1	−29.2	4.3	0.81	436	0.24
35.0	301.9	13.5	−30.1	4.0	0.84	69	0.04
30.0	312.0	13.7	−26.9	4.2	0.84	1248	0.68
25.0	315.4	12.7	−25.1	4.4	0.80	1511	0.83
20.0	319.0	9.3	−21.4	4.9	0.63	1479	0.81
15.0	318.7	5.1	−5.3	5.8	0.28	1194	0.65
10.0	312.9	4.9	52.2	5.0	0.33	794	0.43
5.0	302.6	2.4	17.0	2.9	0.22	495	0.27
0.0	302.2	0.4	−41.0	2.5	0.02	254	0.14
−5.0	304.3	1.5	−82.5	3.0	0.12	290	0.16
−10.0	305.4	3.5	−61.0	3.5	0.32	494	0.27
−15.0	310.0	5.3	−99.8	3.9	0.46	849	0.47
−20.0	313.8	7.7	−112.6	4.8	0.56	950	0.52
−25.0	315.0	11.5	−7.6	4.7	0.74	923	0.51
−30.0	311.9	14.1	166.2	4.6	0.82	716	0.39
−35.0	304.8	12.7	161.2	4.4	0.79	199	0.11

are over the tropics. North of the tropics, Theta is between 19.7 and 35.4 days and reflects the latency of the climate system (the LST peaks about a month after the irradiation maximum). On the southern hemisphere, Theta is about half a year. Since it is defined to be between −182 days and 182 days, the lowest negative and highest positive values are in fact close to each other. The white areas indicate regions that have a time lag compared to the solar radiation, while for the dark areas LST peaks before the solar maximum. At about 10° North and South, there is a transition zone where the Theta parameter shows a noisy and chaotic behavior.

The RMSE (Figure 5.10d) is lowest at the equator (2.5 K) and especially high at 15° N and 75° N (both 5.8 K). The average of all latitudes is 4.5 K. While from the modelling perspective RMSE is an error, it is likewise a measure for scatter of the single observations around the curve and thus the inter-diurnal and inter-annual variations. For this reason it exhibits a similar gradient as YAST. For assessing the model performance the coefficient of determination R^2 (Figure 5.10e) is better suited. To that end, in Figure 5.10e the R^2 of the fitted model is presented. This ACP parameter was first evaluated in this study and shows a very strong latitudinal variation, with an explained variance of 88% at 65° N and only 2% at the Equator.

Finally, in Figure 5.10f the NCSA parameter is presented, which is the number of MODIS LST acquisitions. NCSA is highest in arid areas with peaks at 25° N (1511 acquisitions; 83%) and at 20° S (950 acquisitions; 52%). In the tropics, the NCSA is much lower (254 days; 14%) and at mid-latitudes it is in between (e.g., 645 acquisitions; 35% at 50 °N). This agrees with the climatic cloud occurrence of these macro-climates. However, it can be seen that this is superimposed by orbital effects, which appear as artifacts in the desert regions. This occurs because the orbits have larger overlap at higher latitudes and the probability that a pixel is produced increases with the number of overpasses. In the tropics, pixels above water are more frequently produced.

To further assess the latitudinal ACP3 model performance, the R^2 of the fit as a function of latitudes is presented in Figure 5.11h. The dotted line indicates the 0.75 R^2 boundary, which is considered here as the threshold for defining good model performance. In general, the ACP3 model performs above this threshold between 20° N and 80° N, but its performance deteriorates over the tropics (20° S to 20° N).

The reason for this is the temporal pattern of solar irradiation at different latitudes (Figure 11a–g), which can be considered as primary forcing for the annual cycle in LST. In detail, Figure 11a–g show the potential global radiation (W/m²) at different days of year (left to right) and times of day (top to bottom) for selected latitudes. The values are approximated from the sun's elevation and distance and do not consider atmospheric transfer (atmospheric transmittance τ was set to 0.75). At the North and South Poles (Figure 11a and g, respectively), the potential global radiation is entirely dominated by the annual cycle, while at mid-latitudes (e.g., Figure 11b and f) it is determined by both the annual and the diurnal cycles, and the annual

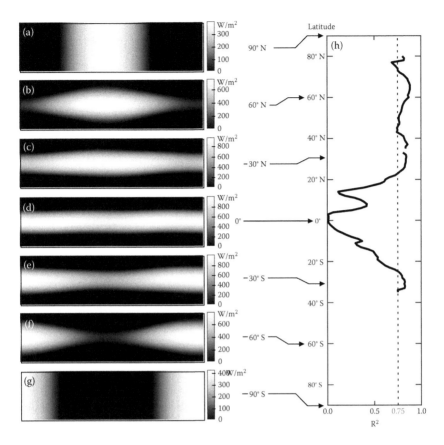

FIGURE 5.11

(a–g) Idealized temporal pattern of global radiation (W/m²) for different latitudes. (h) Coefficient of determination R^2 for Aqua daytime ACP (2011–2015, ACP3) by latitude.

cycle nicely follows a harmonic function. In contrast to the Poles, the Equator is dominated by the diurnal cycle and the annual cycle is comparably flat (Figure 5.11d). Nevertheless, two global radiation maxima corresponding to the spring and fall equinox are observable. The flatness of the annual cycle is also the cause for the poor performance of the ACP3 model (it fits the data with a single sine function) over this area.

To address this issue, an alternative ACP model (i.e., the ACP5 model) that uses a second harmonic function is proposed here. The conceptual suitability is demonstrated in a conceptual test in Figure 5.12. In this experiment an artificial time series was generated by adding white noise to the annual cycle of the potential 13:30 local solar time global radiation and the following latitudes were examined: (a) the Equator, (b) 30° N, (c) 60° N, and (d) the North Pole. Next, the generated time series were approximated with the ACP3 model and the new ACP5 model. While both models fit very well for 60° N ($R^2 = 97\%$ for ACP3% and 98% for ACP5), the difference at the equator is

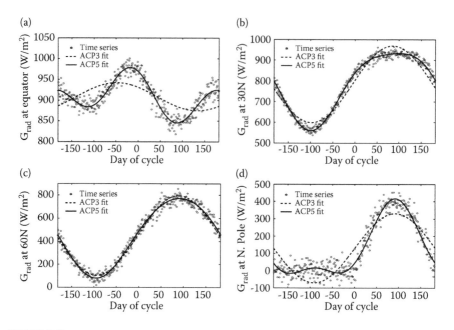

FIGURE 5.12
Fit of ACP3 and ACP5 for simulated global radiation data for (a) Equator, (b) 30° N, (c) 60° N, and (d) North Pole.

tremendous (33% versus 86%). This is also the case for the North Pole, where the ACP5 model results in a substantial improvement (from 79% of ACP3 model to 93%). Even the arctic winter is also comparably well represented, but there is a small secondary maximum. This is since, as for a Fourier Series, more frequencies would be needed to represent the low flat plateau without irradiation.

It is expected, that the R^2 estimated from actual LST data will differ considerably from this conceptual test. This is because the added noise is artificial, of different magnitude, and not necessarily realistic since the magnitude of the fluctuations certainly varies with latitude. In addition, negative radiation obviously does not occur in nature. Nevertheless, this test efficiently demonstrated the suitability of the new ACP5 model to approximate global radiation at very high and very low latitudes and thus it is expected to also improve the fit in LST for these areas.

In addition to the above, Figure 5.13 presents some preliminary results using ACP5 for ROI 2 (Figure 5.1). The boxplots show the increase in R^2 of the ACP5 model compared to the ACP3 model by latitude. As suggested by the previous conceptual test solely considering the radiative forcing, the model performance increases considerably in the tropics (on average 0.09 increase of R^2 between 22.5° N and 22.5° S), while for the mid-latitudes only minor improvements were found (average 0.02). Interestingly, the largest improvement (average increase of 0.21 in R^2) was found at 5° S. While the average improved, the

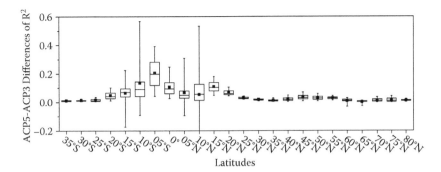

FIGURE 5.13
The difference between the ACP5 R^2 and the ACP3 R^2 for tiles h19v01-h19v12.

variation is high and some pixels showed poorer model results with ACP5. This might be partly due to the preliminary implementation of the ACP5 fit, which is not completely numerically stable yet. However, an especially large variation was found at 10° N, which coincides with the Sahel zone, where the climate is dominated by inter-annual variation rather than seasons. While the ACP5 needs further investigation, the results suggest that they might be a useful alternative especially for the tropics.

5.4 Applications

Various applications for ACP have been suggested, including the study of LST climatology (Bechtel, 2015a,b; Weng and Fu, 2014), the enhancement of the spatial resolution of coarse-scale LST data (Bechtel et al., 2012; Sismanidis et al., 2016, 2017; Zhan et al., 2016), understanding the urban thermal environment (Bechtel, 2012; Bechtel and Daneke, 2012; Bechtel, 2015b), classifying land cover and local climate zones (Bechtel and Daneke, 2012; Bechtel et al., 2016), identifying lithology (Bechtel, 2015a,b; Wei et al., 2017), estimating geothermal energy potential (Hein et al., 2016), topo-climatology (Bechtel, 2016), and monitoring spatiotemporal changes (Hao et al., 2015). In the following sections, we discuss the use of the ACPs for climatological SUHI analysis and for statistically downscaling coarse-scale LST image data to finer spatial resolutions.

5.4.1 Climatological SUHI Analysis

SUHI are often analyzed using few or even just a single LST image, which is entirely inappropriate due to the high spatiotemporal variability of LST. To support this argument, Figure 5.14 shows the daytime SUHI of Paris, France, at selected consecutive almost cloud free days (the data are from Aqua MODIS).

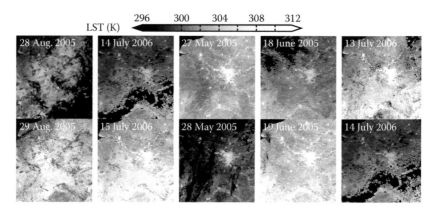

FIGURE 5.14
The Aqua daytime MODIS LST pattern and SUHI of Paris for consecutive days in 2005 and 2006.

Evidently, the outcome of an SUHI analysis that uses only a single LST image can differ dramatically depending on the dates employed. For instance, 14 July 2006 is considerably cooler than both the days before and after (2nd and last column), resulting in different SUHI magnitudes. For the 28th and 29th of August 2005 (first column), not only the mean temperature differs but also the daytime SUHI is hardly recognizable at all. For the 27th of May the daytime SUHI is very strong, while on the 13th of July 2006 the city is warmer than the northern surrounding regions but not the southern. The reasons for such differences comprise atmospheric conditions (radiation, previous cloud cover, etc.), soil conditions (especially soil moisture), phenology, and observation conditions such as viewing time and angle. The latter is relevant since the thermal signal of urban areas is highly anisotropic (Krayenhoff and Voogt, 2016; Lagouarde et al., 2012).

Today, it is increasingly recognized that SUHI analysis and even more importantly, far reaching planning recommendations, cannot be based on arbitrary data selection. Thus, recently more studies employed multi-temporal LST for SUHI analysis (Azevedo et al., 2016; Bahi et al., 2016; Fu and Weng, 2016; Haashemi et al., 2016; Taheri Shahraiyni et al., 2016).

An alternative approach discussed in (Bechtel, 2012, 2015a,b; Fu and Weng, 2015; Huang et al., 2016; Quan et al., 2016; Weng and Fu, 2014) is the use of the ACPs which enables robust SUHI estimates from a large number of satellite acquisitions. This approach either uses the MAST parameter or virtual annual LST patterns derived from the ACP models. Figure 5.15 in particular shows the SUHI of Central European cities as presented in the night time (a) and daytime (b) MYD11A1 MAST (ACP3, years 2011 to 2015). This figure clearly shows that the mean annual SUHI intensity (SUHII) is generally higher at daytime and varies between the cities.

This allows consistent large scale comparison of SUHIIs for different cities, which otherwise could hardly ever be observed under the same or at least

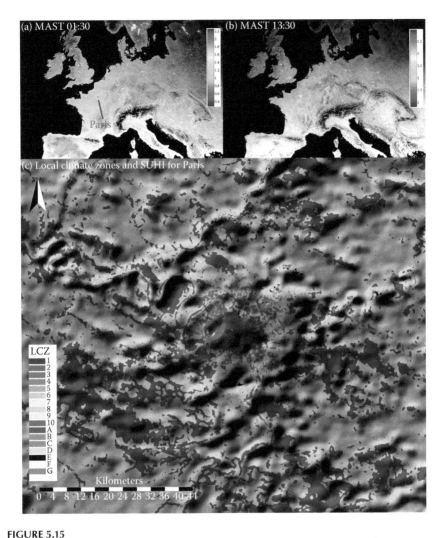

FIGURE 5.15
(See color insert.) Mean annual SUHI at (a) night time and (b) daytime for Central Europe. (c) The nocturnal annual mean SUHI of Paris overlaid as relief on a Local Climate Zone land cover classification according to Stewart and Oke (2012). LCZ 2 = compact midrise (urban core); LCZ 6 = open low rise (suburban housing); LCZ A = dense trees (forest).

comparable conditions. Accordingly, Figure 5.16 shows the distribution of the SUHII for the ROI 1A (Figure 5.1) for the four MODIS overpass times based on 646 identified cities. Generally, the SUHII is again larger at daytime than at night time. Between 10:30 and 13:30, there is a 0.45 K increase in the average SUHII due to the more rapid surface heating of urban materials in this time period. Likewise, there is a cooling between 22:30 and 01:30 of 0.2 K on average.

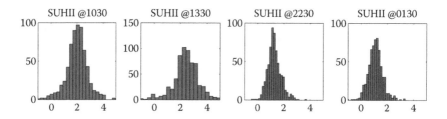

FIGURE 5.16
SUHII for four different overpass times 10:30 (Terra day), 13:30 (Aqua day), 22:30 (Terra night), and 01:30 (Aqua night).

Interestingly, the Aqua and Terra daytime SUHII for the 646 cities are strongly correlated (R = 97% ***) and so are the night time SUHII (R = 90% ***). However, the day and night time SUHIIs are not significantly correlated (despite 10:30 and 01:30, and have a weak negative correlation, R = −13% **). This indicates that day and night time SUHIs have different influencing factors. These influencing factors can now be assessed with this novel SUHI analysis method. For instance, a highly significant correlation of the size of the urban area with the SUHII was found for the Aqua night time MAST SUHII (R = 16% ***) and Terra night time SUHII (R = 17% ***), but not for the respective daytime SUHIIs. We assume that the surrounding land cover and its radiative and thermal properties have strong influence on daytime LST and thus the night time SUHI pattern is expected to show a purer urban footprint. This is of particular relevance, since previously, the vast majority of such studies were based on daytime data due to the limited availability of night time Landsat acquisitions. However, the influencing factors clearly need additional investigation, which are envisaged for the near future.

Additionally, the ACP-based SUHII analysis enables an entirely new characterization and understanding of the temporal characteristics of SUHIs. Figure 5.17 presents the annual course of the day and night time SUHII for Brussels, Berlin, and Paris based on the annual cycle models for the urban and surrounding rural pixels (MYD11A1 ACP3, 2011–2015). The size of the marker indicates the day of year running from 1st of January to 31st of December. First, it can be noted that the modelled SUHII is a sum of harmonic functions and hence is a harmonic function itself. Further, some common characteristics can be seen. In particular, for all three cities the SUHII is higher by day than by night on average. However, this does not apply throughout the year and for all three cities. In detail, in parts of the winter, the night time SUHII is stronger, because the daytime SUHII has a much stronger variation over the year. Additionally, there is a phase shift between daytime and night time, which is stronger for Brussels and Berlin than Paris. Likewise, the average SUHII and the annual variation are lowest for Paris, higher for Berlin, and highest for Brussels.

This strong modelled seasonal variation of the SUHII again underlines the inappropriateness of an SUHI analysis based on single acquisitions. We

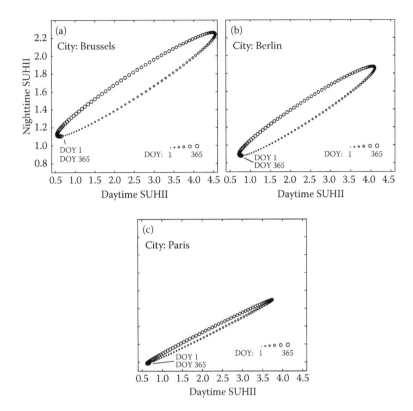

FIGURE 5.17

Modelled surface urban heat island intensity (SUHII) based on the annual cycle of the ACP for Brussels (a), Berlin (b) and Paris (c).

argue that according to the ellipses in Figure 5.17, at least six parameters are needed to quantify the annual characteristic of SUHIs. However, further investigations regarding the fit with observed individual SUHI patterns will be conducted.

Another new paradigm in urban heat island studies that deserves mention is the use of Local Climate Zones (LCZ) (Stewart and Oke, 2012) as descriptors of the urban landscape instead of just the dichotomy between urban and rural. This scheme has been introduced to enhance comparability of observational urban heat island studies, and has been adopted for urban land cover mapping (Bechtel and Daneke, 2012; Bechtel et al., 2015) in the World Urban Database And Portal Tools (WUDAPT) project (See et al., 2015). More recently, it has been uptaken by SUHI studies (Chen et al., 2017; Geletič et al., 2016). Figure 5.15c shows an example of the LCZ of Paris overlaid by an analytical hillshading of the Aqua night time MAST. The light orange colors indicate open low rise, which is a suburban residential morphology, while the red colors indicate compact urban morphologies. The SUHI, represented by the relief, is clearly visible with a cliff at the urban edge and additional temperature increases towards the center, while local cool islands such as parks can be seen as sinks. We argue that like for observational canopy layer UHI studies, the LCZ concept can add much needed detail to the rural-urban dichotomy for SUHI studies. Moreover, the combination with ACP offers a robust and straightforward approach, which allows comparison between different biophysical regions and deserves further investigation in future studies.

5.4.2 Using the ACPs as Disaggregation Kernels for Downscaling LST Image Data

The downscaling of LST is a new branch of TIR remote sensing and has received considerable attention in recent years (Zhan et al., 2013). The enhancement of its spatial resolution offers the ability to extend the use of geostationary LST data to further scientific investigations that have special requirements regarding the data's spatial and temporal resolutions (Sismanidis et al., 2015). This is particularly important because LST changes rapidly both spatially and temporally and the currently available satellite instruments cannot acquire LST data with a spatial and temporal resolution that matches the characteristic scale of these variations. This is mainly due to the trade-off between the spatial and temporal resolution imposed by physical and orbital limitation. Accordingly, remote sensors can acquire either high spatial and low temporal resolution data (e.g., semimonthly acquisitions at ~100 m), or moderate spatial and temporal resolution data (e.g., two images per day with a pixel size of ~1 km), or low spatial and high temporal resolution data (e.g., quarter-hourly acquisitions with a pixel size larger than 3 km) (Bechtel et al., 2012; Sismanidis et al., 2015). To that end, downscaling of high temporal and coarse spatial resolution LST data (i.e., TIR data from geostationary sensors) is a way to address this problem (Bechtel et al., 2012; Inamdar et al., 2008; Keramitsoglou et al. 2013; Sismanidis et al., 2016, 2017; Zakšek and Oštir, 2012; Zhan et al., 2013).

In essence, LST downscaling is a scaling process that aims to enhance the spatial resolution of coarse-scale LST imagery using fine-scale auxiliary datasets (Sismanidis et al., 2015). These auxiliary data (referred to henceforth as disaggregation kernels) are statistically correlated to the LST and available at the desired fine-scale spatial resolution. In general, the workflow of a statistical LST downscaling process consists of three major steps: (i) the upscaling and coregistration of the fine-scale disaggregation kernels to the coarse-scale LST data; (ii) the establishment of a relationship between these two data using an empirical model; and (iii) the application of this relationship to the fine-scale disaggregation kernels so as to generate the downscaled LST (DLST) data. The most employed models are linear or nonlinear regressions depending on the study area. The employed regression strategy can be global or local and the data employed as disaggregation kernels frequently include topography, land cover, vegetation indices, and emissivity data to name just a few.

In recent years, ACP data have also been used as disaggregation kernels with success. In particular, Bechtel et al. (2012, 2013) used ACPs derived from Landsat 7 ETM+ to downscale a ~3 km SEVIRI LST image down to 100 m, while Sismanidis et al. (2016) used MODIS ACPs to downscale a three month long time series of 3 km/15 min SEVIRI LST data to 1 km/15 min. Furthermore, Sismanidis et al. (2017) also provided evidence that the use of multi-temporal ACPs as disaggregation kernels can improve the retrieval of the diurnal LST range from downscaled LST data.

Generally, the use of ACPs as disaggregation kernels offers many advantages: (i) being derived from TIR data, they can explain much of the LST spatial variation; (ii) they can also provide information about how this location-specific variability (e.g., the effects of topographic shading) changes with time when multi-temporal ACPs are employed (e.g., ACP that correspond to various times within a day, such as morning, noon or night hours). This information is especially useful when downscaling diurnal LST time series (Sismanidis et al., 2016, 2017); (iii) they perform consistently over various land cover types, landscapes, time of day, and climatic conditions; and (iv) they can help limit the number of LST disaggregation kernels, which is more practical and performs better as some studies suggest (Bechtel et al., 2012; Bisquert et al., 2016). In particular, the consistent performance of the ACPs as disaggregation kernels over various land cover types and landscapes ensures the applicability of the downscaling algorithms in various landscapes and climatic zones and avoids limitation to the area for which they were developed, as is the case today.

The advantages of using the ACPs as disaggregation kernels are further highlighted in Figure 5.18. This figure presents the change of spatial distribution and magnitude of the DLST RMSE using different disaggregation kernels based on the data and methods used in Sismanidis et al. (2016): the DLST RMSE was estimated from a three month long time series of DLST data retrieved from the Spinning Enhanced Visible and Infrared Imager (SEVIRI) and downscaled to 1 km, while the reference data used were the corresponding 1 km MOD11A1 and MYD11A1 collection-5 daily LST. The first line of this figure (noted as scheme 1 in Figure 5.18) corresponds to the DLST RMSE using MAST, YAST, and Theta (retrieved from a 5-year time series of collection-5 MODIS LST data using the ACP3 method) as disaggregation kernels. The second line (noted as scheme 2a in Figure 5.18) corresponds to the DLST RMSE of a large set of disaggregation kernels including altitude, slope, aspect, land cover, vegetation indices, and emissivity data (presented in detail in Sismanidis et al., 2016), and the third line (noted as scheme 2b in Figure 5.18) to the joint set of all kernels. To ensure comparability, all other parameters were set the same, that is, the regression tool used (a support vector regression machine), the downscaling algorithm, and the scenes used for estimating the RMSE.

The comparison of the DLST RMSE maps of Figure 5.18 reveals that the ACPs (first line) achieved the smallest RMSE for each time examined and that the distribution of the RMSE hotspots is more uniform than for the other two schemes. Furthermore, the inclusion of MAST, YAST, and Theta considerably improved the performance of the downscaling process especially for the daytime data. This is evident through the comparison of schemes 2a and 2b,

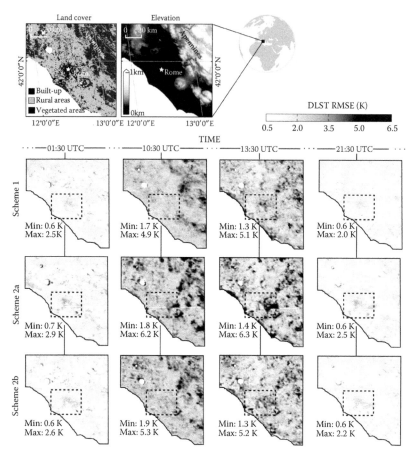

FIGURE 5.18
DLST RMSE improvement by including ACPs as LST disaggregation kernels: Scheme 1 uses only MAST, YAST, and Theta; Scheme 2a uses altitude, slope, aspect, land cover, vegetation indices, and emissivity data as disaggregation kernels; and Scheme 2b all the above. The square corresponds to the city of Rome, Italy.

where the dark hotspots of the maps of the second line, which correspond to the mountainous regions located at the right and the urban area at the center of each map, were reduced noticeably. For the night time data, the improvement is less pronounced but a certain change from dark to brighter tones is still observable. These results suggest that the location and time-specific information that the ACPs provide can substantially improve the downscaling accuracy and facilitate the downscaling of diurnal LST data. The higher RMSE values over the Apennines are partially due to pronounced directional effects.

Furthermore, ACPs have been used in other types of methods for generating high spatiotemporal LST, such as fusion methods that combine data from various satellites, for example, Landsat and MODIS. In particular, Zhan et al. (2016) used an ACP model in conjunction with a diurnal temperature cycle

(DTC) model to indirectly generate DLST by controlling their parameters, and Weng and Fu (2014) used the ACPs in the fusion model they proposed for generating daily MODIS LST at Landsat resolution.

5.5 Conclusions

ACPs are a robust estimate of the annual cycle of LST and provide a cloud-gap free representation of thermal surface characteristics. They can be used to derive virtual LST patterns for every day of year, which represent a long term average surface temperature climatology without short term variations.

ACPs have been computed globally using the MODIS 1 km LST products before. In this chapter, they were first produced using the latest collection (v6) of these MODIS data. The parameters MAST, YAST, and NCSA were most affected by the algorithm refinements of the new collection. The greatest differences were found over the Sahara desert in North Africa, which is due to an improvement of the LST acquisitions algorithm over sand and bare soil. The acquisition over cities was improved resulting in substantially improved data coverage as shown for Paris, France. However, despite missing acquisitions over most of the year, the ACPs were still estimated impressively well from MODIS collection 5, which underlines their robustness against data gaps.

The coefficient of determination R^2 was implemented as an additional parameter and confirmed a comparably bad performance of the simple ACP model (ACP3) in the tropics. Thus an alternative model with a second harmonic function (ACP5) was proposed here, which resulted in a considerable improvement of the fit in the tropics.

ACPs have a large number of potential applications. Here, SUHI analysis and LST downscaling were presented as examples. It was found that the ACPs allow a more comprehensive view of the SUHI and its annual course, which should become more widely used in future studies. In particular, it was found that SUHI analysis based on single acquisitions are largely random due to inter-diurnal differences and strong seasonal variations. ACPs allow SUHI comparison between many cities under comparable conditions as well as statistical analysis of the influencing factors.

References

Azevedo, J. A., Chapman, L. and C. L. Muller. 2016. Quantifying the daytime and night-time urban heat island in Birmingham, UK: A comparison of satellite derived land surface temperature and high resolution air temperature observations. *Remote Sensing* 8(2): 153. doi: 10.3390/rs8020153.

Bahi, H., Rhinane, H., Bensalmia, A., Fehrenbach, U. and D. Scherer. 2016. Effects of urbanization and seasonal cycle on the surface urban heat island patterns in the coastal growing cities: A case study of Casablanca, Morocco. *Remote Sensing* 8(10): 829.

Bechtel, B. 2011. Multitemporal Landsat data for urban heat island assessment and classification of local climate zones. *Urban Remote Sensing Event (JURSE), 2011 Joint.* IEEE: 129–132. doi: 10.1109/JURSE.2011.5764736. http://ieeexplore.ieee.org/abstract/document/5764736/.

Bechtel, B. 2012. Robustness of annual cycle parameters to characterize the urban thermal landscapes. *Geoscience and Remote Sensing Letters, IEEE* 9(5): 876–880. doi: 10.1109/LGRS.2012.2185034.

Bechtel, B. 2015a. A new global climatology of annual land surface temperature. *Remote Sensing* 7(3): 2850–2870. doi: 10.3390/rs70302850.

Bechtel, B. 2015b. Die Hitze in der Stadt verstehen – wie sich die jahreszeitliche Temperaturdynamik von Städten aus dem All beobachten lässt. In *Globale Urbanisierung—Perspektive aus dem All*, Taubenböck, H., Wurm, M., Esch T. and Dech, S. 205–216. Berlin Heidelberg: Springer Spektrum.

Bechtel, B., Alexander, P. J., Böhner, J., Ching, J., Conrad, O., Feddema, J., Mills, G., See, L. and I. Stewart. 2015. Mapping local climate zones for a worldwide database of the form and function of cities. *ISPRS International Journal of Geo-Information* 4(1): 199–219.

Bechtel, B., Boehner, J., Zaksek, K. and S. Wiesner. 2013. Downscaling of diurnal land surface temperature cycles for urban heat island monitoring. In *Urban Remote Sensing Event (JURSE), 2013 Joint*, 091–094. https://doi.org/10.1109/JURSE.2013.6550673.

Bechtel, B. 2016. The climate of the Canary islands by annual cycle parameters. *ISPRS—International Archives of the Photogrammetry, Remote Sensing and Spatial Information Sciences XLI-B8 (June)* 243–250. doi: 10.5194/isprs-archives-XLI-B8-243-2016.

Bechtel, B. and C. Daneke. 2012. Classification of local climate zones based on multiple earth observation data. *IEEE Journal of Selected Topics in Applied Earth Observations and Remote Sensing* 5(4): 1191–1202. doi: 10.1109/JSTARS.2012.2189873.

Bechtel, B., See, L., Mills, G. and M. Foley. 2016. Classification of local climate zones using SAR and multispectral data in an arid environment. *IEEE Journal of Selected Topics in Applied Earth Observations and Remote Sensing* 9(7), 3097–3105. doi: 10.1109/JSTARS.2016.2531420.

Bechtel, B., Zaksek, K. and G. Hoshyaripour. 2012. Downscaling land surface temperature in an Urban area: A case study for hamburg, Germany. *Remote Sensing* 4(10): 3184–3200. doi: 10.3390/rs4103184.

Becker, F. and Z.-L. Li. 1995. Surface temperature and emissivity at various scales: Definition, measurement and related problems. *Remote Sensing Reviews* 12, 225–253.

Bisquert, M., Sánchez, J. M. and V. Caselles. 2016. Evaluation of disaggregation methods for downscaling MODIS land surface temperature to landsat spatial resolution in barrax test site. *IEEE Journal of Selected Topics in Applied Earth Observations and Remote Sensing* 9(4), 1430–1438.

Bobrowski, M., Bechtel, B., Oldeland, J., Weidinger, J., Schickhoff, U. In review. Improving ecological niche models with remotely sensed data for predicting the current distribution of Betula Utilis in the Himalayan region. *Progress in Physical Geography.*

Chen, Yu-Cheng, Tzu-Ping Lin, and Wan-Yu Shih. 2017. Modeling the Urban Thermal Environment Distributions in Taipei Basin Using Local Climate Zone (LCZ). In *Urban Remote Sensing Event (JURSE), 2017 Joint, 1–4. IEEE*. Dubai, United Arab Emirates. http://jurse2017.com/index.html.

Friedl, M. A., Sulla-Menashe, D., Tan, B., Schneider, A., Ramankutty, N., Sibley, A. and X. Huang. 2010. MODIS Collection 5 global land cover: Algorithm refinements and characterization of new datasets. *Remote Sensing of Environment* 114(1): 168–182. doi: 10.1016/j.rse.2009.08.016.

Fu, P. and Q. Weng. 2015. Temporal dynamics of land surface temperature from landsat TIR time series images. *IEEE Geoscience and Remote Sensing Letters* 12(10): 2175–2179. doi: 10.1109/LGRS.2015.2455019.

Fu, P. and Q. Weng. 2016. A time series analysis of urbanization induced land use and land cover change and its impact on land surface temperature with Landsat imagery. *Remote Sensing of Environment* 175, 205–214.

Geletič, J., Lehnert, M. and P. Dobrovolný. 2016. Land surface temperature differences within local climate zones, based on two central european cities. *Remote Sensing* 8(10), 788.

Gillespie, A., Rokugawa, S., Matsunaga, T., Cothern, J. S., Hook, S. and A. B. Kahle. 1998. A temperature and emissivity separation algorithm for Advanced Spaceborne Thermal Emission and Reflection Radiometer (ASTER) images. *IEEE Transactions on. Geoscience and Remote Sensing* 36(4), 1113–1126, doi: 10.1109/36.700995.

Guillevic, P. C., Privette, J. L., Coudert, B., Palecki, M. A., Demarty, J., Ottlé, C. and Augustine, J. A. 2012. Land Surface Temperature Product Validation Using NOAA's Surface Climate Observation Networks-Scaling Methodology for the Visible Infrared Imager Radiometer Suite (VIIRS). *Remote Sensing of Environment* 124, 282–298.

Haashemi, S., Weng, Q., Darvishi, A. and S. K. Alavipanah. 2016. Seasonal variations of the surface urban heat island in a semi-arid city. *Remote Sensing* 8(4): 352. doi: 10.3390/rs8040352.

Hao, P., Niu, Z., Zhan, Y., Wu, Y., Wang, L. and Y. Liu. 2015. Spatiotemporal changes of urban impervious surface area and land surface temperature in Beijing from 1990 to 2014. *GIScience and Remote Sensing*, 53(1), 63–84.

Hein, P., Ke, Z., Anke, B., Olaf, K., Zhonghe, P. and S. Haibing. 2016. Quantification of exploitable shallow geothermal energy by using borehole heat exchanger coupled ground source heat pump systems. *Energy Conversion and Management* 127: 80–89.

Huang, F., Zhan, W., Voogt, J., Hu, L., Wang, Z., Quan, J., Weimin, J. and Z. Guo. 2016. Temporal upscaling of surface urban heat island by incorporating an annual temperature cycle model: A tale of two cities. *Remote Sensing of Environment* 186: 1–12.

Inamdar, A. K., French, A., Hook, S., Vaughan, G. and W. Luckett. 2008. Land surface temperature retrieval at high spatial and temporal resolutions over the southwestern United States. *Journal of Geophysical Research* 113(D7): D07107.

Jimenez-Munoz, J. C. and J. A. Sobrino. 2010. A single-channel algorithm for land-surface temperature retrieval from ASTER data. *IEEE Geoscience and Remote Sensing Letters* 7(1): 176–179. doi: 10.1109/LGRS.2009.2029534.

Keramitsoglou, I., Kiranoudis, C. T., Ceriola, G., Weng, Q. and U. Rajasekar. 2011. Identification and analysis of urban surface temperature patterns in Greater Athens, Greece, using MODIS imagery. *Remote Sensing of Environment* 115(12): 3080–3090.

Keramitsoglou, I., Daglis I. A., Amiridis, V., Chrysoulakis, N., Ceriola, G., Manunta, P., Maiheu, B., De Ridder, K., Lauwaet, D. and M. Paganini. 2012. Evaluation of satellite-derived products for the characterization of the urban thermal environment. *Journal of Applied Remote Sensing* 6(1): 061704–061704. doi: 10.1117/1. JRS.6.061704.

Keramitsoglou, I., Kiranoudis, C. T. and Q. Weng. 2013. Downscaling geostationary land surface temperature imagery for Urban analysis. *IEEE Geoscience and Remote Sensing Letters* 10(5): 1253–1257.

Klein, I., Gessner, U., Dietz, A. J. and C. Kuenzer. 2017. Global WaterPack—A 250 m resolution dataset revealing the daily dynamics of global inland water bodies. *Remote Sensing of Environment* 198: 345–362. doi: 10.1016/j. rse.2017.06.045

Krayenhoff, E. S. and J. A. Voogt. 2016. Daytime thermal anisotropy of urban neighbourhoods: morphological causation. *Remote Sensing* 8(2): 108. doi: 10.3390/ rs8020108.

Lagouarde, J. P., Hénon, A., Irvine, M., Voogt, J., Pigeon, G., Moreau, P., Masson V. and P. Mestayer. 2012. Experimental characterization and modelling of the nighttime directional anisotropy of thermal infrared measurements over an urban area: Case study of Toulouse (France). *Remote Sensing of Environment* 117: 19–33.

Li, Z.-L., Tang, B.-H., Wu, H., Ren, H., Yan, G., Wan, Z., Trigo, I. F. and J.A. Sobrino. 2013a. Satellite-derived land surface temperature: Current status and perspectives. *Remote Sensing of Environment* 131: 14–37. doi: 10.1016/j.rse.2012.12.008.

Li, Z.-L., Wu, H., Wang, N., Qiu, S., Sobrino, J. A., Wan, Z., Tang, B.-H. and G. Yan. 2013b. Land surface emissivity retrieval from satellite data. *International Journal of Remote Sensing* 34: 3084–3127. doi: 10.1080/01431161.2012.716540

Mitraka, Z., Chrysoulakis, N., Kamarianakis, Y., Partsinevelos, P. and A. Tsouchlaraki. 2012. Improving the estimation of Urban surface emissivity based on sub-pixel classification of high resolution satellite imagery. *Remote Sensing of Environment* 117, 125–134.

Norman, J.M. and F. Becker. 1995. Terminology in thermal infrared remote sensing of natural surfaces. *Agricultural and Forest Meteorology* 77(3–4), 153–166.

Price, J. C. 1984. Land surface temperature measurements from the split window channels of the NOAA 7 Advanced Very High Resolution Radiometer. *Journal of Geophysical Research: Atmospheres* 89(D5), 7231–7237.

Qin, Z., Karnieli, A. and P. Berliner. 2001. A mono-window algorithm for retrieving land surface temperature from Landsat TM data and its application to the Israel-Egypt border region. *International Journal of Remote Sensing* 22(18): 3719–3746. doi: 10.1080/01431160010006971.

Quan, J., Zhan, W., Chen, Y., Wang, M. and L. Wang. 2016. Time series decomposition of remotely sensed land surface temperature and investigation of trends and seasonal variations in surface urban heat islands. *Journal of Geophysical Research: Atmospheres* 121(6), 2015JD024354. doi: 10.1002/2015JD024354.

Schowengerdt, R.A. 2006. *Remote Sensing: Models and Methods for Image Processing*. Academic press. London UK.

See, L. et al. 2015. Developing a Community-Based Worldwide Urban Morphology and Materials Database (WUDAPT) Using Remote Sensing and Crowdsourcing for Improved Urban Climate Modelling. In *Urban Remote Sensing Event (JURSE), 2015 Joint, 1–4.* doi: 10.1109/JURSE.2015.7120501.

Sismanidis, P., Keramitsoglou, I., Bechtel, B. and C. T. Kiranoudis. 2017. Improving the downscaling of diurnal land surface temperatures using the annual cycle parameters as disaggregation dernels. *Remote Sensing* 9(1), 23.

Sismanidis, P., Keramitsoglou, I. and C. T. Kiranoudis. 2015. A satellite-based system for continuous monitoring of Surface Urban Heat Islands. *Urban Climate* 14, 141–153.

Sismanidis, P., Keramitsoglou, I., Kiranoudis, C. T. and B. Bechtel. 2016. Assessing the capability of a downscaled urban land surface temperature time series to reproduce the spatiotemporal features of the original data. *Remote Sensing* 8(4): 274.

Snyder, W. C., Wan, Z., Zhang, Y. and Y. Z. Feng. 1998. Classification-based emissivity for land surface temperature measurement from space. *International Journal of Remote Sensing* 19: 2753–2774.

Sobrino, J. A., Jiménez-Muñoz, J. C. and L. Paolini. 2004. Land surface temperature retrieval from LANDSAT TM 5. *Remote Sensing of Environment* 90(4): 434–440. doi: 10.1016/j.rse.2004.02.003.

Sobrino, J. A., Jiménez-Muñoz, J. C., Sòria, G., Romaguera, M., Guanter, L., Moreno, J., Plaza, A. and P. Martínez. 2008. Land surface emissivity retrieval from different VNIR and TIR sensors. *IEEE Transactions on Geoscience and Remote Sensing* 46(2): 316–327.

Sobrino, J. A., Li, Z. L., Stoll, M. P. and F. Becker. 1996. Multi-channel and multi-angle algorithms for estimating sea and land surface temperature with ATSR data. *International Journal of Remote Sensing* 17(11): 2089–2114. doi: 10.1080/01431169608948760.

Stewart, I. D. and T. Oke. 2012. Local Climate Zones for Urban Temperature Studies. *Bulletin of the American Meteorological Society* 93(12), 1879–1900.

Taheri Shahraiyni, H., Sodoudi, S., El-Zafarany, A., Abou El Seoud, T., Ashraf, H. and K. Krone. 2016. A Comprehensive Statistical Study on Daytime Surface Urban Heat Island during Summer in Urban Areas, Case Study: Cairo and Its New Towns. *Remote Sensing* 8(8): 643.

Townshend, J. R. G., Justice, C. O., Skole, D., Malingreau, J.-P., Cihlar, J., Teillet, P., Sadowski, F. and S. Ruttenberg. 1994. The 1 Km Resolution Global Data Set: Needs of the International Geosphere Biosphere Programme. *International Journal of Remote Sensing* 15, 3417–3441. doi: 10.1080/01431169408954338

Yang, J., Wong, M. S., Menenti, M. and J. Nichol. 2015. Study of the geometry effect on land surface temperature retrieval in urban environment, ISPRS J. *Photogrammetry and Remote Sensing* 109, 77–87, doi: 10.1016/j.isprsjprs.2015.09.001.

Valor, E. and V. Caselles. 1996. Mapping land surface emissivity from NDVI: Application to European, African, and South American areas. *Remote Sensing of Environment* 57(3): 167–184. doi: 10.1016/0034-4257(96)00039-9.

Voogt, J. A. and T. R. Oke. 2003. Thermal remote sensing of urban climates. *Remote Sensing of Environment* 86: 370–384.

Voogt, J. A. and T. R. Oke. 1998. Effects of urban surface geometry on remotely-sensed surface temperature. *Int. J. Remote Sens.* 19(5), 895–920, doi: 10.1080/014311698215784.

Wan, Z. 2007. Collection-5 MODIS land surface temperature products users guide. ICESS, University of California, Santa Barbara. http://icess.eri.ucsb.edu/modis/LstUsrGuide/MODIS_LST_products_Users_guide_Collection-6.pdf (accessed: December 28, 2014).

Wan, Z. 2008. New refinements and validation of the MODIS land-surface temperature/emissivity products. *Remote Sensing of Environment* 112(1), 59–74.

Wan, Z. 2014. New refinements and validation of the collection-6 MODIS land-surface temperature/emissivity product. *Remote Sensing of Environment* 140: 36–45. doi: 10.1016/j.rse.2013.08.027.

Wan, Z. 2015. MOD11A1 MODIS/Terra Land Surface Temperature/Emissivity Daily L3 Global 1 km SIN Grid V006. NASA EOSDIS Land Processes DAAC. https://doi.org/10.5067/modis/mod11a1.006

Wan, Z. and Dozier, J. 1996. A generalized split-window algorithm for retrieving land-surface temperature from space. *IEEE Transactions on Geoscience and Remote Sensing* 34(4), 892–905. doi: 10.1109/36.508406.

Wan, Z., Zhang, Y., Zhang, Q. and Z. L. Li. 2004. Quality assessment and validation of the MODIS global land surface temperature. *International Journal of Remote Sensing* 25(1), 261–274. doi: 10.1080/0143116031000116417.

Wang, K. and S. Liang. 2009. Evaluation of ASTER and MODIS land surface temperature and emissivity products using long-term surface longwave radiation observations at SURFRAD sites. *Remote Sensing of Environment* 113(7), 1556–1565.

Wei, Jiali, Xiangnan, Liu, Chao, Ding, Meiling, Liu, Ming, Jin, and Dongdong, Li. 2017. Developing a Thermal Characteristic Index for Lithology Identification Using Thermal Infrared Remote Sensing Data. *Advances in Space Research* 59(1), 74–87.

Weng, Q. 2009. Thermal infrared remote sensing for urban climate and environmental studies: Methods, applications, and trends. *ISPRS Journal of Photogrammetry and Remote Sensing* 64(4): 335–344.

Weng, Q. and P. Fu. 2014. Modeling annual parameters of clear-sky land surface temperature variations and evaluating the impact of cloud cover using time series of Landsat TIR data. *Remote Sensing of Environment* 140, 267–278. doi: 10.1016/j.rse.2013.09.002.

Zakšek, K. and K. Oštir. 2012. Downscaling land surface temperature for urban heat island diurnal cycle analysis. *Remote Sensing of Environment* 117, 114–124. doi: 10.1016/j.rse.2011.05.027.

Zhan, W. et al. 2013. Disaggregation of remotely sensed land surface temperature: Literature survey, taxonomy, issues, and caveats. *Remote Sensing of Environment* 131(19), 119–139.

Zhan, W. et al. 2016. Disaggregation of remotely sensed land surface temperature: A new dynamic methodology. *Journal of Geophysical Research: Atmospheres* 121, 1–17.

6

Impervious Surface Estimation by Integrated Use of Landsat and MODIS Time Series in Wuhan, China

Zhang Lei and Qihao Weng

CONTENTS

6.1 Introduction 121
6.2 Study Area 123
6.3 Methodology 123
    6.3.1 Data Preprocessing 124
    6.3.2 Reconstruction of Time Series BCI 125
    6.3.3 Similarity of Temporal Features 126
    6.3.4 Classification Based on Semi-Supervised SVM 126
6.4 Results and Discussion 127
    6.4.1 Annual Dynamics of Impervious Surfaces.............................. 127
    6.4.2 Classification Accuracy 128
6.5 Conclusions.. 131
Acknowledgments 131
References 131

6.1 Introduction

Dynamics of impervious surfaces has a significant influence on studies of urban environments, including biogeochemical cycles, climate change, and biodiversity. Monitoring the dynamics of impervious surfaces is helpful for urban planning and management, especially for sustainable cities. Remote sensing technology provides spatially consistent data with fine spatial resolution and high temporal frequency for city-level studies on impervious surfaces over a long time period. Numerous approaches have been proposed to map dynamics of impervious surfaces using multi-temporal remote sensing imagery, such as Defense Meteorological Satellite Program's Operational Linescan System (DMSP/OLS) night time light data (Zhang & Seto, 2011; Liu et al., 2012; Ma et al., 2012), Landsat archive (Bhatta, 2009; Ahmed & Ahmed, 2012; Bagan, & Yamagata, 2012), MODIS imagery (Mertes et al., 2015), and multi-sensor imagery (Pandey, Joshi, & Seto, 2013; Shao & Liu, 2014).

Previous methods mainly focused on spectral differences to identify impervious surfaces from multi-temporal imagery. These methods fell into three categories: per-pixel analysis, sub-pixel analysis, and object-based analysis. Per-pixel methods included classification methods at the pixel level (Yin et al., 2011; Vermeiren et al., 2012), spectral indices (Villa, 2012), thresholding technique (Liu et al., 2012; Ma et al., 2012), image fusion (Taubenböck et al., 2012), and so on. Sub-pixel methods, such as spectral mixture analysis (Lu, Moran, & Hetrick, 2011; Michishita, Jiang, Z, & Xu 2012; Deng & Wu, 2013), regression analysis (Xian & Crane, 2005; Sexton et al., 2013), and machine learning algorithms (Du, Xia, & Feng, 2015) were helpful to extract urban area abundance within each pixel. Object-based methods were suited to extract impervious surfaces from high resolution (HR) data (Durieux, Lagabrielle, & Nelson, 2008; Jacquin, Misakova, & Gay, 2008; Dupuy, Barbe, & Balestrat, 2012). Additionally, Samal & Gedam (2015) used the object-based method in moderate spatial resolution imagery to classify urban area consistently in time series image analysis. Sun et al. (2013) used object-oriented classification in combination with spatial metrics to improve the understanding of impervious surfaces processes from Landsat imagery. Existing methods usually required single classification of all the stacked images and paid little attention to the spatiotemporal changes of various land covers before classification. Therefore, most methods cannot effectively handle the mixed pixel problem in coarse or medium spatial resolution imagery or the intra-class spectral variability problem in high spatial resolution imagery.

Temporal domain has shown its advantages in solving class confusion between classes with similar spectral characteristics (Schneider, 2012). Specifically, time series Landsat data is helpful when distinguishing vegetation change and impervious surface change due to phenological change, inter-annual climatic variability, and long term change trends (Bhandari et al., 2012). Time series Landsat data at fine spatial resolution and high temporal frequency has been successfully applied for characterizing dynamics of impervious surfaces (Sexton et al., 2013; Li, Gong, & Liang, 2015). Mapping impervious surfaces using time series of Landsat data is proving to be very useful for assessing and monitoring urban growth (Powell et al., 2008; Gao et al., 2012; Zhou et al., 2012; Zhang et al., 2013).

Most previous methods used time series Landsat data to check classification consistency to correct illogical class transitions. However, time series Landsat data is still underutilized because temporal features have been barely introduced to identify different land covers. This study aims at extracting temporal features from time series Landsat data to identify the differences between impervious and pervious surfaces. The objectives of this study are: (1) to reconstruct time series Biophysical Composition Index (BCI); (2) to measure similarity of temporal features of impervious and pervious surfaces from reconstructed time series BCI; and (3) to classify temporal features using semi-supervised Support Vector Machine (SVM) and map dynamics of impervious surfaces at an annual frequency, over the case study of Wuhan, China, from 2000 to 2015.

6.2 Study Area

Wuhan city, as the capital of Hubei Province, China, is located in the eastern Jianghan Plain, between 29°58′–31°22′ N and 113°41′–115°05′ E, covering an area of 8494 km². It has a humid subtropical climate with abundant rainfall. The monthly 24-hour average temperature ranges from 3.7°C in January to 28.7°C in July, the annual mean temperature is 16.63°C, and the mean annual precipitation is about 1,269 mm.

Wuhan City, as one of the largest metropolitan areas in China, has witnessed a rapid urbanization since 2000. Its population was recorded at 10 million in 2015, increasing from 7.49 million in 2000. For the period from 2000 to 2014, the Gross Domestic Product (GDP) of Wuhan increased almost eight-fold, from 120 billion yuan in 2000 (approximately US $19 billion) to 1006 billion yuan (approximately US $162 billion) in 2014 (Wuhan Statistics Bureau). In the past 16 years, Wuhan's per capita GDP has grown from 16,206 yuan in 2000 to 98,527 yuan (China City Statistical Yearbook).

In this study, 168 Landsat Surface Reflectance Climate Data Record (Landsat CDR) images were ordered and downloaded from the USGS Earth Explorer (Worldwide Reference System (WRS): WRS-2, Path: 123, Row: 39) in level L1T for Landsat TM, ETM+, and Operational Land Imager (OLI). Only images with cloud cover less than 20% were included in the analyses. Image subsets from Landsat imagery with an area of 7024.56 square kilometers were used to monitor dynamics of impervious surfaces in Wuhan. The geographic location of the study area is shown in Figure 6.1.

6.3 Methodology

To map annual dynamics of impervious surfaces, temporal features were developed to differentiate impervious and pervious surfaces using reconstructed time series BCI data from time series Landsat imagery. Dynamic time warping (DTW) distance was employed to measure the similarity of temporal features. Finally, SVM algorithm was implemented to map annual impervious surfaces based on temporal features. The procedures for mapping annual dynamics of impervious surfaces in Wuhan city are as follows (Figure 6.2).

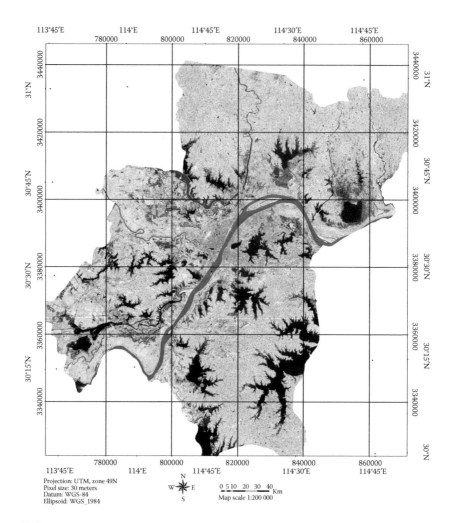

FIGURE 6.1
(**See color insert.**) The geographic location of Wuhan city.

6.3.1 Data Preprocessing

All the Landsat L1T data were registered to the 1984 World Geodetic System Universal Transverse Mercator (WGS-84 UTM) Zone 49 North projection system and resampled at 30-meter spatial resolution. Landsat 4–5 Thematic Mapper (TM) and Landsat 7 Enhanced Thematic Mapper Plus (ETM+) surface reflectance data are generated from the Landsat Ecosystem Disturbance Adaptive Processing System (LEDAPS) method (Masek et al., 2006), which applied MODIS atmospheric correction routines to Landsat L1 data products. Landsat 8 surface reflectance (L8SR) data are generated from the L8SR algorithm (USGS, 2015). Cloud, cloud shadow, and snow mask were calculated

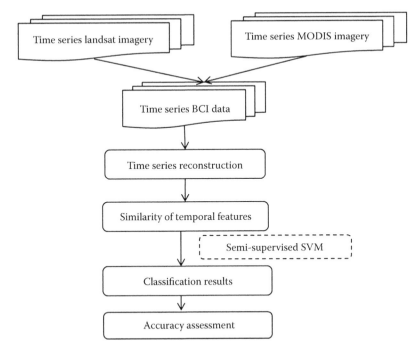

FIGURE 6.2
Procedures for mapping annual dynamics of impervious surfaces in Wuhan city.

using the Fmask algorithm for all scenes (Zhu & Woodcock, 2012). The locations of scan-line corrector (SLC)-off data were identified using band-specific gap mask files in each SLC-off data product. All masked pixels were converted to NaN (not a number) values. To address the problem of missing data, Flexible Spatiotemporal Data Fusion (FSDAF) (Zhu et al., 2016) was conducted to predict the missing values by fusing Landsat time series and time-adjacent MODIS time series in the year with all cloud contaminated images. The cloud free Landsat image in a time adjacent year was used as the base image for prediction.

6.3.2 Reconstruction of Time Series BCI

Biophysical Composition Index (BCI) (Deng & Wu, 2012) is aimed at identifying different urban biophysical compositions. It has been demonstrated to be effective in identifying the characteristics of impervious surfaces and vegetation, and distinguishing bare soil from impervious surfaces. BCI, involving Tasseled Cap (TC) transformation and vegetation-impervious surface-soil (VIS) triangle model, was given as:

$$BCI = \frac{(TC_1 + TC_3)/2 - TC_2}{(TC_1 + TC_3)/2 + TC_2} \tag{6.1}$$

where TC_1 was "high albedo," TC_3 was "low albedo," and TC_2 was "vegetation." TC_i ($i = 1, 2$, and 3) were three normalized TC components. Each derived TC component was then linearly normalized to the range from 0 to 1.

Normalized difference water index (NDWI) (Gao, 1996) was applied to mask water bodies.

$$NDWI = \frac{G - NIR}{G + NIR} \tag{6.2}$$

where G represented reflectance values in the green band and NIR represented reflectance values in the near infrared band. $NDWI$ produced an image in which water bodies typically had positive values, while nonwater pixels typically had negative values. $NDWI$ had a native scaling of -1 to $+1$.

Since available Landsat images were unevenly distributed in time due to cloud contamination, time series BCI were necessary to be reconstructed. Given time series $T = \{T_1, T_2, \cdots \in, T_t\}$, t was year number, T_1 represented BCI images in the first year, and T_i represented BCI images in the ith year, $1 \leq i \leq t$. Images within each year would be clustered into four parts: spring (February-April), summer (May-July), autumn (August-October), and winter (November-January). That is, $T_i = \{T_i^{sp}, T_i^{su}, T_i^{au}, T_i^{wi}\}$, where T_i^{sp}, T_i^{su}, T_i^{au}, and T_i^{wi}, were mean values of each part, respectively. T_i^{sp} represented BCI images in the spring of the ith year. Thus, the dimension of reconstructed time series was calculated by multiplying total year number by four ($16 \times 4 = 64$).

6.3.3 Similarity of Temporal Features

In this study, stable training samples were defined as samples which experienced no change over time. Training samples were collected by visual interpretation on Landsat images with the help of high spatial resolution images from Google Earth. Three land cover categories, impervious surfaces, pervious surfaces, and water, were included. Since water bodies were masked using NDWI, this study focused on the differences between impervious and pervious surfaces.

Dynamic time warping (DTW) distance was introduced in the proposed method to measure the similarity of temporal features, because DTW has been shown to be effective in finding the optimal match between two time series and in evaluating similarity based on their shapes (Jeong, Jeong, & Omitaomu, 2011). The pseudo code for calculating DTW distance between two time series can be found in Petitjean, Ketterlin, and Gançarski (2011). However, it was difficult for DTW distance to calculate the average of a set of time series. The averaging method in Petitjean, Ketterlin, and Gançarski (2011) was applied to calculate the average of the time series.

6.3.4 Classification Based on Semi-Supervised SVM

Support Vector Machine (SVM) has been successfully employed in the classification field due to its good generalization ability with limited training samples and globally optimal performance with high dimensions (Burges, 1998). Given time series data in the period from year t_1 to year t_m, temporal features in the period from t_1 to t_i would be used to obtain impervious surfaces' distribution in year t_i, $1 \leq i \leq m$, using an SVM classifier.

Take year t_i for example, the procedure of semi-supervised SVM algorithm was shown as follows. First, input data was $X = \left\{ X^l_{m \times n \times d}, X^u_{m \times n \times d} \right\}$, l represented training samples, and u represented testing samples. m and n were the length and width of the study area, d was the dimension of the temporal feature for each pixel. Next, SVM was conducted to train X^l to get classifier C1 using polynomial kernel and classifier C2 using radial basis function. Parameters of polynomial kernels and radial basis function were determined by a grid search method using a cross validation approach. After that, X^u was forecast by classifiers C1 and C2, and forecasting results R1 and R2 were obtained. Then, R1 and R2 were compared to get pixels with high creditability and put these pixels into X^l to update X^l. Finally, if all X^u were classified, exit loop.

6.4 Results and Discussion

6.4.1 Annual Dynamics of Impervious Surfaces

Taking one pixel for example, the reconstruction of time series BCI was shown in Figure 6.3. Reconstructed time series BCI reduced the temporal dimension of original time series BCI, and provided regular temporal information for each pixel. The annual maps of impervious surfaces in Wuhan city are shown in Figure 6.4. Light gray represented impervious surfaces, dark gray represented pervious surfaces, and medium gray represented water. During the 16 years between 2000 and 2015, Wuhan city has experienced a rapid exponential urban expansion. Annual impervious surfaces in the study area were shown in Figure 6.5. Impervious surfaces of Wuhan city expanded from 515.27 km² in 2000 to 2308.91 km² in 2015 at an annual average rate of 21.76%.

Over the past 16 years, impervious surfaces' expansion of Wuhan city in the different periods was different.

1. During the period from 2000 to 2002, impervious surfaces grew slowly with an annual average rate of 6.93%. Impervious surfaces increased from 515.27 km² in 2000 to 586.74 km² in 2002. New impervious surfaces were built in urban areas.

2. The period from 2003 to 2007 witnessed the first rapid impervious surfaces expansion in Wuhan city with an annual average rate of 9.91%. Impervious surfaces increased from 609.95 km² in 2003 to 912.12 km² in 2007. Impervious surfaces expanded along the Changjiang River and Hanjiang River in this period.

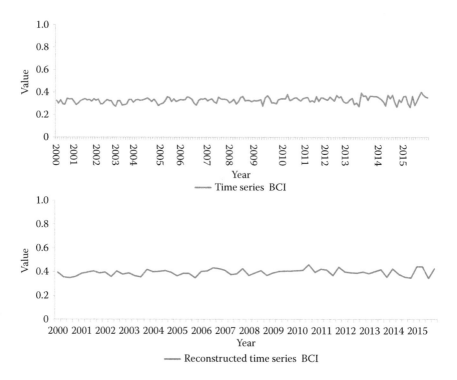

FIGURE 6.3
The reconstruction of time series BCI.

3. For the period from 2008 to 2011, Wuhan city experienced another accelerating urban sprawl with an annual average rate of 16.26%. Impervious surfaces increased from 1004.83 km^2 in 2008 to 1658.29 km^2 in 2011. Impervious surfaces expanded from urban areas to rural areas.

4. During the period from 2012 to 2015, the annual average growth rate dropped to 6.06%. Impervious surfaces increased from 1858.21 km^2 in 2012 to 2308.91 km^2 in 2015. Most of urban expansion occurred around the urban core area.

6.4.2 Classification Accuracy

Confusion matrices were used to assess classification accuracy in this study. The overall accuracy, producer's accuracy, and user's accuracy of impervious surfaces maps were shown in Figure 6.6. "P-" represented producer's accuracy and "U-" represented user's accuracy. The overall accuracies for the period from 2000 to 2015 ranged from 76.42% in 2000 to 90.71% in 2015. The average overall accuracy in this study was 86.16%.

In this study, image numbers, cloud cover, and SLC-off data determined the dimension of temporal features, which influenced the performance

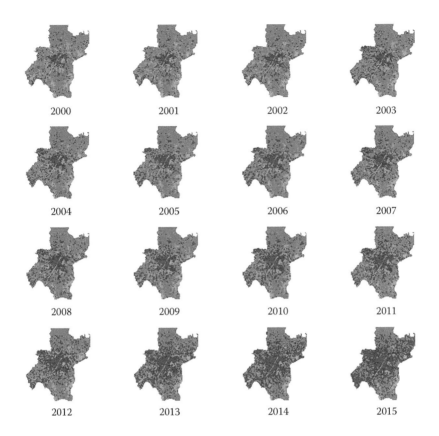

FIGURE 6.4
(See color insert.) Annual maps of impervious surfaces in Wuhan city.

of SVM classifier and had an impact on classification accuracy. In 2000, classification accuracy was the lowest (76.42%) since this year only had five available images. Image numbers contributed to increase the dimension of temporal features over time and enhanced the separability between impervious and pervious surfaces. For example, although image numbers and cloud cover in 2001 (twelve images with average cloud cover 4.28%) and 2002 (ten images with average cloud cover 4.36%) were similar, the classification accuracy in 2002 (84.11%) was higher than classification accuracy in 2001 (77.64%) because the dimension of temporal features in 2002 was larger than that in 2001.

Additionally, we used semi-supervised SVM with the same parameters to classify least cloud-contaminated Landsat images directly for each year to make a comparison with the proposed method. The classification accuracies were achieved using the same training and testing samples. The overall accuracy, producer's accuracy, and user's accuracy of impervious surfaces maps were shown in Figure 6.7.

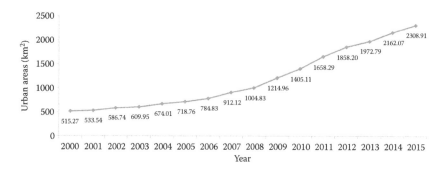

FIGURE 6.5
Annual impervious surfaces in Wuhan city.

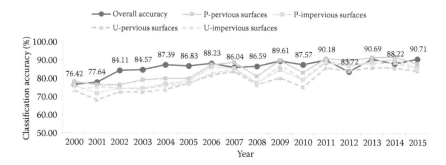

FIGURE 6.6
The overall accuracy, producer's accuracy, and user's accuracy of annual impervious surfaces maps using the proposed method.

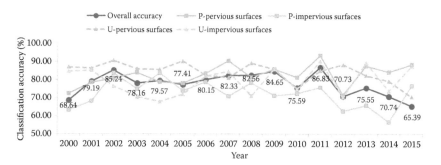

FIGURE 6.7
The overall accuracy, producer's accuracy, and user's accuracy of annual impervious surfaces maps using a semi-supervised SVM.

The classification accuracies using semi-supervised SVM, with an average overall accuracy of 77.67%, were lower than that using the proposed method. The comparison results presented in Figures 6.6 and 6.7 showed that the proposed method using temporal features produced a higher level of classification accuracy than did the SVM classifier only using spectral differences in identifying impervious surfaces.

6.5 Conclusions

The significance of this study lies in the proposal of a new reliable methodological approach for measuring similarity of temporal features to map dynamics of impervious surfaces at an annual frequency using time series Landsat data. Reconstructed time series BCI helped to identify the spectral differences between impervious and pervious surfaces. Similarity of temporal features measured by DTW contributed to explore the temporal differences between impervious and pervious surfaces. Finally, SVM was used to classify impervious surfaces based on temporal features. Due to the high temporal dimension of temporal information and the difficulty for obtaining ground truth data, this study applied a semi-supervised SVM method to temporal features. The methodology has demonstrated good performance over the case study of Wuhan city in China, from 2000 to 2015. The average classification accuracy of 86.16% that tested for the case study verified the effectiveness of our method. The results would be meaningful for studies on urban development and planning. In our next study, more Landsat images with cloud contamination and data gaps from SLC-off data will be used to improve the temporal resolution of temporal features and explore the values of missing data in time series analysis of land cover.

Acknowledgments

This work was supported by Postdoctoral Innovative Talent Support Plan (BX201700175).

References

Ahmed, B., & Ahmed, R. 2012. Modeling urban land cover growth dynamics using multi-temporal satellite images: A case study of Dhaka, Bangladesh. *ISPRS International Journal of Geo-Information*, 1(1), 3–31.

Bagan, H., & Yamagata, Y. 2012. Landsat analysis of urban growth: How Tokyo became the world's largest megacity during the last 40years. *Remote Sensing of Environment*, 127, 210–222.

Bhandari, S., Phinn, S., & Gill, T. 2012. Preparing Landsat Image Time Series (LITS) for monitoring changes in vegetation phenology in Queensland, Australia. *Remote Sensing*, 4(6), 1856–1886.

Bhatta, B. 2009. Analysis of urban growth pattern using remote sensing and GIS: A case study of Kolkata, India. *International Journal of Remote Sensing*, 30(18), 4733–4746.

Burges, C. J. 1998. A tutorial on support vector machines for pattern recognition. *Data Mining and Knowledge Discovery*, 2(2), 121–167.

Deng, C., & Wu, C. 2012. BCI: A biophysical composition index for remote sensing of urban environments. *Remote Sensing of Environment*, 127, 247–259.

Deng, C., & Wu, C. 2013. A spatially adaptive spectral mixture analysis for mapping subpixel urban impervious surface distribution. *Remote Sensing of Environment*, 133, 62–70.

Du, P., Xia, J., & Feng, L. 2015. Monitoring urban impervious surface area change using China-Brazil Earth Resources Satellites and HJ-1 remote sensing images. *Journal of Applied Remote Sensing*, 9(1), 096094–096094.

Dupuy, S., Barbe, E., & Balestrat, M. 2012. An object-based image analysis method for monitoring land conversion by artificial sprawl use of RapidEye and IRS data. *Remote Sensing*, 4(2), 404–423.

Durieux, L., Lagabrielle, E., & Nelson, A. 2008. A method for monitoring building construction in urban sprawl areas using object-based analysis of Spot 5 images and existing GIS data. *ISPRS Journal of Photogrammetry and Remote Sensing*, 63(4), 399–408.

Gao, B. C. 1996. NDWI—a normalized difference water index for remote sensing of vegetation liquid water from space. *Remote Sensing of Environment*, 58(3), 257–266.

Gao, F., de Colstoun, E. B., Ma, R., Weng, Q., Masek, J. G., Chen, J., ... & Song, C. 2012. Mapping impervious surface expansion using medium-resolution satellite image time series: A case study in the Yangtze River Delta, China. *International Journal of Remote Sensing*, 33(24), 7609–7628.

Jacquin, A., Misakova, L., & Gay, M. 2008. A hybrid object-based classification approach for mapping urban sprawl in periurban environment. *Landscape and Urban Planning*, 84(2), 152–165.

Jeong, Y. S., Jeong, M. K., & Omitaomu, O. A. 2011. Weighted dynamic time warping for time series classification. *Pattern Recognition*, 44(9), 2231–2240.

Li, X., Gong, P., & Liang, L. 2015. A 30-year (1984–2013) record of annual urban dynamics of Beijing City derived from Landsat data. *Remote Sensing of Environment*, 166, 78–90.

Liu, Z., He, C., Zhang, Q., Huang, Q., & Yang, Y. 2012. Extracting the dynamics of urban expansion in China using DMSP-OLS nighttime light data from 1992 to 2008. *Landscape and Urban Planning*, 106(1), 62–72.

Lu, D., Moran, E., & Hetrick, S. 2011. Detection of impervious surface change with multitemporal Landsat images in an urban–rural frontier. *ISPRS Journal of Photogrammetry and Remote Sensing*, 66(3), 298–306.

Ma, T., Zhou, C., Pei, T., Haynie, S., & Fan, J. 2012. Quantitative estimation of urbanization dynamics using time series of DMSP/OLS nighttime light data: A comparative case study from China's cities. *Remote Sensing of Environment*, 124, 99–107.

Masek, J. G., Vermote, E. F., Saleous, N. E., Wolfe, R., Hall, F. G., Huemmrich, K. F., ... & Lim, T. K. 2006. A Landsat surface reflectance dataset for North America, 1990–2000. *IEEE Geoscience and Remote Sensing Letters*, 3(1), 68–72.

Mertes, C. M., Schneider, A., Sulla-Menashe, D., Tatem, A. J., & Tan, B. 2015. Detecting change in urban areas at continental scales with MODIS data. *Remote Sensing of Environment*, 158, 331–347.

Michishita, R., Jiang, Z., & Xu, B. 2012. Monitoring two decades of urbanization in the Poyang Lake area, China through spectral unmixing. *Remote Sensing of Environment*, 117, 3–18.

Pandey, B., Joshi, P. K., & Seto, K. C. 2013. Monitoring urbanization dynamics in India using DMSP/OLS night time lights and SPOT-VGT data. *International Journal of Applied Earth Observation and Geoinformation*, 23, 49–61.

Petitjean, F., Ketterlin, A., & Gançarski, P. 2011. A global averaging method for dynamic time warping, with applications to clustering. *Pattern Recognition*, 44(3), 678–693.

Powell, S. L., Cohen, W. B., Yang, Z., Pierce, J. D., & Alberti, M. 2008. Quantification of impervious surface in the Snohomish water resources inventory area of western Washington from 1972–2006. *Remote Sensing of Environment*, 112(4), 1895–1908.

Samal, D. R., & Gedam, S. S. 2015. Monitoring land use changes associated with urbanization: An object based image analysis approach. *European Journal of Remote Sensing*, 48, 85–99.

Schneider, A. 2012. Monitoring land cover change in urban and peri-urban areas using dense time stacks of Landsat satellite data and a data mining approach. *Remote Sensing of Environment*, 124, 689–704.

Sexton, J. O., Song, X. P., Huang, C., Channan, S., Baker, M. E., & Townshend, J. R. 2013. Urban growth of the Washington, DC–Baltimore, MD metropolitan region from 1984 to 2010 by annual, Landsat-based estimates of impervious cover. *Remote Sensing of Environment*, 129, 42–53.

Shao, Z., & Liu, C. 2014. The integrated use of DMSP-OLS nighttime light and MODIS data for monitoring large-scale impervious surface dynamics: A case study in the Yangtze River Delta. *Remote Sensing*, 6(10), 9359–9378.

Sun, C., Wu, Z. F., Lv, Z. Q., Yao, N., & Wei, J. B. 2013. Quantifying different types of urban growth and the change dynamic in Guangzhou using multi-temporal remote sensing data. *International Journal of Applied Earth Observation and Geoinformation*, 21, 409–417.

Taubenböck, H., Esch, T., Felbier, A., Wiesner, M., Roth, A., & Dech, S. 2012. Monitoring urbanization in mega cities from space. *Remote Sensing of Environment*, 117, 162–176.

USGS. 2015. Product Guide: Provisional Landsat 8 Surface Reflectance Product (Version 1.7, September 2015). United States Geological Service. (https://landsat.usgs.g ov/sites/default/files/documents/lasrc_product_guide.pdf)

Vermeiren, K., Van Rompaey, A., Loopmans, M., Serwajja, E., & Mukwaya, P. 2012. Urban growth of Kampala, Uganda: Pattern analysis and scenario development. *Landscape and Urban Planning*, 106(2), 199–206.

Villa, P. 2012. Mapping urban growth using soil and vegetation index and landsat data: The Milan (Italy) city area case study. *Landscape and Urban Planning*, 107(3), 245–254.

Xian, G., & Crane, M. 2005. Assessments of urban growth in the Tampa Bay watershed using remote sensing data. *Remote Sensing of Environment*, 97(2), 203–215.

Yin, J., Yin, Z., Zhong, H., Xu, S., Hu, X., Wang, J., & Wu, J. 2011. Monitoring urban expansion and land use/land cover changes of Shanghai metropolitan area during the transitional economy (1979–2009) in China. *Environmental Monitoring and Assessment*, 177(1–4), 609–621.

Zhang, Q., & Seto, K. C. 2011. Mapping urbanization dynamics at regional and global scales using multi-temporal DMSP/OLS nighttime light data. *Remote Sensing of Environment*, 115(9), 2320–2329.

Zhang, X., Pan, D., Chen, J., Zhan, Y., & Mao, Z. 2013. Using long time series of Landsat data to monitor impervious surface dynamics: A case study in the Zhoushan Islands. *Journal of Applied Remote Sensing*, 7(1), 073515–073515.

Zhou, B., He, H. S., Nigh, T. A., & Schulz, J. H. 2012. Mapping and analyzing change of impervious surface for two decades using multi-temporal Landsat imagery in Missouri. *International Journal of Applied Earth Observation and Geoinformation*, 18, 195–206.

Zhu, X., Helmer, E. H., Gao, F., Liu, D., Chen, J., & Lefsky, M. A. 2016. A flexible spatiotemporal method for fusing satellite images with different resolutions. *Remote Sensing of Environment*, 172, 165–177.

Zhu, Z., & Woodcock, C. E. 2012. Object-based cloud and cloud shadow detection in Landsat imagery. *Remote Sensing of Environment*, 118, 83–94.

Part III

Time Series Image Applications

7

Mapping Land Cover Trajectories Using Monthly MODIS Time Series from 2001 to 2010

Shanshan Cai and Desheng Liu

CONTENTS

7.1 Introduction 137
7.2 Study Area and Data 139
    7.2.1 Study Area 139
    7.2.2 Image Data 140
    7.2.3 Reference Data... 140
7.3 Methods... 141
    7.3.1 Algorithm Overview 141
    7.3.2 Detecting Change Dates 141
    7.3.3 Generating Adaptive Time Series...................................... 142
    7.3.4 Modified SVM Classification 143
    7.3.5 Integrated Training and Classification 143
    7.3.6 Trajectory Reconstruction.. 144
    7.3.7 Comparison of Adaptive Time Series with Full Length
         Time Series 144
    7.3.8 Accuracy Assessment... 144
7.4 Results 145
    7.4.1 Trajectory Mapping Results... 145
    7.4.2 Accuracy Assessment of Adaptive Classification Results 146
    7.4.3 Accuracy Assessment of Trajectory Mapping Results............. 148
    7.4.4 Comparison of Adaptive Time Series with Full Length
         Time Series... 149
7.5 Discussion... 150
7.6 Conclusions... 152
References... 153

7.1 Introduction

Land cover determines critical surface conditions associated with many environmental and ecological processes at various spatial and temporal scales (Meyer et al. 1994). Changes in land cover have a wide range of impacts on the Earth systems, including climate change, loss of biodiversity, and soil degradation (Turner et al. 1990; Vitousek et al. 1997; Lambin et al. 1999; Foley et al. 2005). With the widespread natural and anthropogenic land changes, the need for monitoring land cover change on a regular basis over large areas has never been greater (Rindfuss et al. 2004; Steffen et al. 2004; Liu & Cai 2012). Satellite remote sensing plays an essential role in mapping and monitoring of land cover change with its continuous observations of the land surface (Alves & Skole 1996; Defries & Belward 2000; Coppin et al. 2004).

Moderate Resolution Imaging Spectroradiometer (MODIS) has provided global satellite imagery of the land surface at a 1-2 day repeat frequency since 2000. A number of MODIS land products have been generated to support global change studies (Justice et al. 2002), among which the Collection 5 MODIS Land Cover Type (MLCT), MCD12Q1, provides annually updated global land cover maps at 500-m spatial resolution from 2001(Friedl et al. 2002, 2010). The MODIS land cover product has been used to provide important land cover properties to a wide variety of regional and global studies (Gerten et al. 2004; Myhre et al. 2005; Singh & Bhaduri 2010; Wiedinmyer et al. 2011; Turner et al. 2012).

The Collection 5 MLCT product is generated from a boosted decision tree classifier algorithm on a calendar year basis from 2001 to 2013. The input features to the annual land cover classification consist of twelve sets of 32-day average nadir bidirectional reflectance distribution function (BRDF) adjusted reflectance data, enhanced vegetation indices (EVI) data, and land surface temperature (LST) supplemented by their annual metrics (minimum, maximum and mean values) (Friedl et al. 2010). The use of a monthly time series over a year allows the algorithm to better distinguish different land cover classes by exploiting vegetation phenology and temporal variability characteristic of land cover types from multi-seasonal images (Friedl et al. 2010). However, the effectiveness of using multi-seasonal images throughout a year relies on the assumption that land cover types do not change within the period of a calendar year. This assumption is hardly valid in reality as land cover change occurs as a continuous process that varies in space and time. When land cover change occurs within a calendar year, the temporal variation of input features not only corresponds to the assumed vegetation phenology and temporal characteristic of the same land cover type, but also reflects the abrupt change in land cover type. Consequently, land cover classification using a time series spanning an entire year can result in inaccurate results when change occurs within the time series period. Moreover, the annual land cover mapping strategy has two limitations: (1) the pixel-specific time of land cover change cannot be determined within each calendar year, and (2) land cover changes lasting less than a year cannot be mapped. Hence, land cover change within a year should not be ignored when considering the benefit of temporal information (Verbesselt et al. 2010). It is necessary to detect the time of land cover change and use it to determine appropriate starting and ending dates of a time series used in land cover classification (Zhu & Woodcock 2014).

In this chapter, we propose an integrated approach to map land cover trajectories using MODIS time series data. In contrast to MLCT product that maps land cover on a calendar year basis, we directly map the land cover trajectories for each pixel by simultaneously analyzing a 10-year time series. Built on our previous work on change date detection (Cai & Liu 2015), our method first segments the continuous time series into numerous sets of adaptive time series corresponding to a stable land cover period and then applies a modified support vector machine (SVM) algorithm to adaptive time series in land cover classification. We demonstrate the method using the MODIS time series between 2001 and 2010 for a study area in southeast Ohio, USA. The land cover trajectories were reconstructed at 32-day intervals where most of the land cover changes that are missed or observed later than actual times in annual mapping practice can be captured.

7.2 Study Area and Data

7.2.1 Study Area

The study area is located in the southern part of Vinton County, Ohio. A subset of 60 × 80 columns/rows (15 km × 20 km) was selected from the MODIS tile h11v05 at 250-m spatial resolution (Figure 7.1). According to previous studies (e.g., Liu & Cai 2012), the land cover change in this area is highly dynamic as

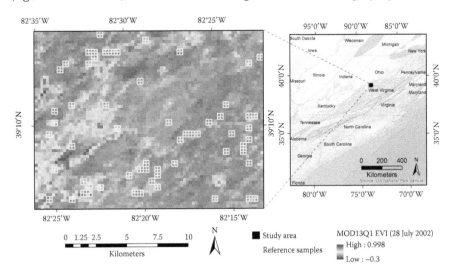

FIGURE 7.1
Location of the study area and the reference samples in MODIS tile h11v05. The image on the left, a subset of MOD13Q1 EVI scene generated for 28 July 2002, illustrates the spatial extent of the study area. The small squares represent the locations of the reference samples.

TABLE 7.1

The Input Features Used for Land Cover Classification

Source	Input Features	Temporal Resolution	Spatial Resolution
MCD43A4	Nadir BRDF-adjusted reflectance band 1, 2, 6	16-day	500 m
MOD13Q1	EVI index	16-day	250 m

a result of human activities such as logging and reforestation, agricultural use and abandonment, and industrial mining and abandonment in the past several decades.

7.2.2 Image Data

The MODIS time series from January 2001 to December 2010 (Table 7.1) is used as input data for land cover classification, including the MODIS nadir BRDF-adjusted reflectance (NBAR) data provided by the MODIS BRDF/albedo product (MCD43A4) at 500-m resolution and the enhanced vegetation index (EVI) data provided by the MOD13Q1 product at 250-m resolution. The MCD43A4 NBAR data are atmospherically corrected and marked for water, clouds, heavy aerosols, and cloud shadows (Schaaf et al. 2002). It was resampled to 250-m resolution with a nearest neighbor method to match the resolution of the EVI data. The EVI data is computed from NBAR surface reflectance and also uses the blue band to remove residual atmosphere contamination caused by smoke and sub-pixel thin clouds (MODIS 1999).

MODIS NBAR and EVI data were originally generated based on 16 days of acquisitions, which were averaged to a 32-day time series. The complete input features for each calendar year include 12 sets of 32-day average NBAR in Bands 1, 2, and 6, and 32-day average EVI data, providing a total of 48 features. The purpose of using fewer spectral bands was to improve computational efficiency. According to the study of Zhou et al. (2013), time series of MODIS reflectance in Bands 1, 2, and 6 are capable of producing similar classification results to those generated from all seven bands.

7.2.3 Reference Data

A set of 244 pixels were randomly selected from the study area to serve as reference data, including both stable and changed land cover trajectories from 2001 to 2010. The land cover trajectory and the dates of changes at each pixel in the dataset were identified based on Landsat data (Cai & Liu 2015). A series of 94 Landsat images for the scene of Path 19 Row 33 were downloaded from the USGS Landsat archive for the period from 2000 to 2011, including Landsat 5 Thematic Mapper (TM) scenes from 2000 to 2011 and Landsat 7 Enhanced Thematic Mapper Plus (ETM+) scenes from 1999 to 2003. Each pixel was assigned to one of six classes (forest, shrub, grass, crop, bare soil, and water) in each 32-day interval by visual inspection. The time of change dates identified from the Landsat time series was converted to the time index for the 32-day interval as accurately as possible.

7.3 Methods

7.3.1 Algorithm Overview

The proposed trajectory mapping approach integrates change detection and land cover classification to track the trajectories of land cover change. Figure 7.2 shows the workflow of the trajectory mapping method. First, the dates of land cover changes are detected for each pixel from MODIS time series. Next, adaptive time series are generated based on the change dates detected. After that, a modified SVM method is applied to land cover classification of adaptive time series. Finally, land cover trajectories for each pixel are reconstructed from the classes identified for the adaptive time series and smoothed within each stable period. The details of the procedures are described in the following sections.

7.3.2 Detecting Change Dates

In our previous study (Cai & Liu 2015), we have developed a sub-annual change detection (SCD) algorithm to identify the change dates for each pixel using a 16-day time series of MODIS Normalized Difference Vegetation Index (NDVI) product MOD13Q1. Figure 7.3 shows the procedure of the SCD algorithm

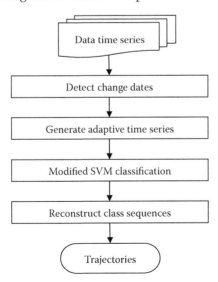

FIGURE 7.2
Flowchart showing the workflow of trajectory mapping.

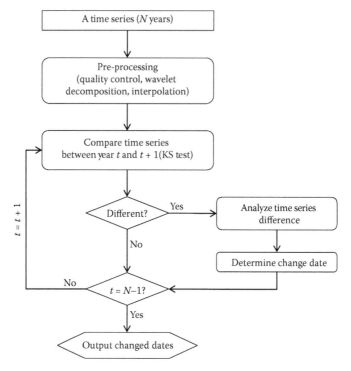

FIGURE 7.3
Flowchart of the sub-annual change detection method. (Adapted from Cai, S., & Liu, D. 2015. *Remote Sensing*, 7, 8705–8727.)

proposed in Cai & Liu (2015). The SCD algorithm starts with pre-processing the NDVI time series with simple smoothing techniques to remove noise and obtain an uninterrupted data stream. It then detects change points within the time series in two hierarchical steps at "year" and "date" levels. In the first step, annual time series are compared to detect years that have significant difference from neighboring years. Within those "changed years", the second step detects the observation points that do not follow seasonal patterns. The final change points represent the dates with abrupt land cover changes significantly different from seasonal variation. Details about the method and the accuracy assessment of the detected change dates are described in Cai & Liu (2015). In this study, we converted the detected change dates at 16-day intervals to the time indices at the 32-day time interval to match our image data.

7.3.3 Generating Adaptive Time Series

Adaptive time series is defined as a subset of annual time series in which land cover type does not change. The rational of generating adaptive time series for land cover classification is to avoid the issues caused by land cover change mentioned in the introduction. Meanwhile, there is still some useful temporal information in adaptive time series that can be used to better differentiate vegetated land cover classes at non-growing seasons. To generate adaptive time series, we adjusted the starting and ending time points of the annual time series according to the detected change dates to ensure no land cover change within each adaptive time series. Specifically, for the years when change dates are detected, the original annual time series were broken into various short time series at the change dates; and for the years without changes, the full length annual time series were kept.

7.3.4 Modified SVM Classification

As the occurrences of land cover changes vary in space and time, the lengths of the generated adaptive time series also vary across different pixels and/ or years. If we consider a time series containing a full year record (i.e., for 12 months) as the standard length, the adaptive time series for the years with land cover changes can be regarded as a time series with absent features after (or before) change dates. In this way, classification with adaptive time series can be treated as classification with absent features with respect to its full year record. Therefore, we adopted a modified Support Vector Machines (SVM) method developed by Chechik et al. (2008) in the classification of adaptive time series.

The modified Support Vector Machines (SVM) method was initially designed for classification with structurally missing features and originally applied in edge prediction in metabolic pathways and automobile detection in natural images (Chechik et al. 2008). It was reformulated from standard max-margin classification to handle non-existing features. The key to this formulation is that kernels must handle absent features correctly. Chechik et al. (2008) indicated that kernel functions whose inputs depend on inner product can be easily transformed into a kernel that handles absent features by considering only features that are valid for both samples and skipping the rest. Polynomial and Sigmoidal kernels are that type of kernel. Chechik et al. (2008) suggested that a second-degree polynomial kernel performed better than other orders, so we adopted the second-degree polynomial kernel for the modified SVM method in this study. Although the radial basis function (RBF) kernel is usually preferred because of its better performance and high computational efficiency than other kernel functions, it is difficult to transform RBF to a kernel that handles absent features properly.

7.3.5 Integrated Training and Classification

The modified SVM classifier was trained in an integrated approach where the reference adaptive time series for the ten years were pooled together for a supervised classification. This integrated strategy has three advantages for the training process. First, the classifier can learn more absent patterns of the reference adaptive time series from all ten years compared to one year. Second, the number of reference adaptive time series for a particular class varied from year to year and certain classes may have very few in some years because land cover changed at some reference pixels over the ten years. If annual classification was performed for each year using the corresponding reference adaptive time series in that year, the accuracy could be largely biased and the result may be unreliable in some years. In the integrated approach, all available reference adaptive time series were used to avoid such problems. Third, the number of reference samples for the integrated training process was ten times that of the original one, which is important for both training and testing of the classifier. With the integrated approach, each adaptive time series was classified to a class no matter in which year it was acquired. Therefore, only one classification was needed to generate the class maps for all ten years.

7.3.6 Trajectory Reconstruction

The land cover trajectory was obtained at each pixel through a reconstruction process based on the classification result. A sequence of land cover classes were first generated based on the time period that an adaptive time series spanned and the class it was classified to. Temporal consistency was then assured for each stable period. For each stable period in which two or more land cover types were generated, a majority vote was used to assign a land cover class to this stable period. The land cover trajectory at each pixel was finally composed of a sequence of land cover classes with the number and time span corresponding to those of stable periods.

7.3.7 Comparison of Adaptive Time Series with Full Length Time Series

Adaptive time series proposed to handle changes is expected to offer better results by representing the major land cover in annual land cover mapping than full length time series (i.e., 12 months) which ignore changes. In order to compare the accuracy of annual land cover mapping by using adaptive time series with full length annual time series, an integrated classification was performed using full length annual time series and the adaptive time series corresponding to the longest time period in each year. A land cover map was then generated from the results of integrated classifications for each instance.

Standard SVM method was used for both full length annual time series and adaptive time series, where absent features in adaptive time series were filled with zeros (Chechik et al. 2008; Salberg & Jenssen 2012). In addition, the modified SVM method was applied to the adaptive time series to better handle absent features. Second-degree polynomial function was used as the kernel function in both standard and the modified SVM.

7.3.8 Accuracy Assessment

The reference pixels were divided into 10 subsets and a 10-fold cross-validation analysis was conducted to compute average values of accuracy indices. Specifically, one subset was retained for testing and the other nine subsets were used for training. This was repeated for 10 times such that each subset was used for testing each time. The means were then calculated for the accuracy indices from the 10 results. Overall accuracy, kappa coefficient, producer's accuracy, and user's accuracy were calculated for each classification.

For trajectory mapping, we calculated two additional indices to evaluate the accuracy of trajectories mapping based on the reference trajectories at the 244 pixels. The first one is *annual trajectory accuracy*, which was calculated for each year as the ratio of the number of pixels with correct trajectories to the total number of reference pixels. The second index is *overall trajectory accuracy* and was calculated as the ratio of the number of pixels with correct trajectories in the entire time period (i.e., ten years in this study) to the total number of reference pixels. For both indices, the trajectory at a pixel over a certain time was considered correct only if the class sequence of the trajectory matches the reference sequence. The annual trajectory accuracy evaluated annual trajectory map, and the overall trajectory accuracy is equivalent to the accuracy of change detection over ten years. The trajectory accuracy also depends on the accuracy of change date detection, especially the accuracy of detected number of changes. To remove the effect of change date detection and demonstrate the errors only from classification and trajectory reconstruction, the reference change dates were used to generate stable period and adaptive time series for the reference pixels.

For the comparison of adaptive time series and full length annual time series, the accuracy of the annual map was calculated for each year as the ratio of pixels with the correct class to the total number of reference pixels. The land cover class of a reference pixel in each year was determined as the major land cover class which occurred in the majority time of that year.

7.4 Results

7.4.1 Trajectory Mapping Results

Figure 7.4 illustrates the trajectory map of 2001–2010 generated by the modified SVM method for the study area. It shows the class maps for 2001 and 2010 and the land cover changes in the middle. For pixels with detected changes, the squares in between the top and bottom plots represent the changed classes at the change points on the time axis, while the blanks represent no land cover change at other time points or for the pixels without any change. Figure 7.4 also demonstrates the spatial and temporal dynamics of land cover change at 32-day resolution, which is not observable in annual land cover maps.

The land cover trajectories illustrated in Figure 7.4 provide an approximation to the continuous land cover process, from which the land cover state at any time point within the time frame of the trajectory can be derived. As an

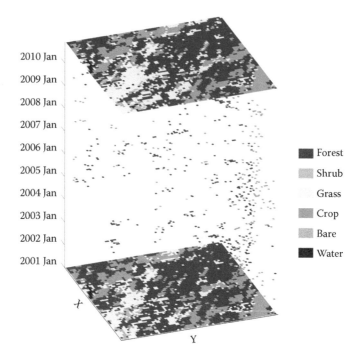

FIGURE 7.4
(**See color insert.**) Land cover trajectories from 2001 to 2010. The plots at the bottom and the top correspond to the class maps for the starting point (January 2001) and ending point (December 2010), respectively. The land cover classes between the two time points are displayed only if land cover changes occurred.

example, a temporal sequence of land cover maps in June of each year from 2001 to 2010 is illustrated in Figure 7.5.

Figure 7.6 illustrates the annual areas (km²) of the various types of land cover change which are more frequent than other types between 2001 and 2010. Note that the change from forest to grass was the dominant land cover change in the study area.

7.4.2 Accuracy Assessment of Adaptive Classification Results

The classification results using the modified SVM method were compared with a standard SVM classification where both RBF and second-degree polynomial functions were applied to show their different strengths and absent features were filled with zeros. Accuracy indices for all classification are summarized in Table 7.2. The modified SVM method performed best followed by the standard SVM with RBF kernel. The modified SVM method generated an overall accuracy of 0.787, which is 0.02 and 0.05 greater than the standard SVM with RBF and polynomial kernels, respectively. The kappa coefficient for the modified SVM is 0.03 and 0.08 greater than

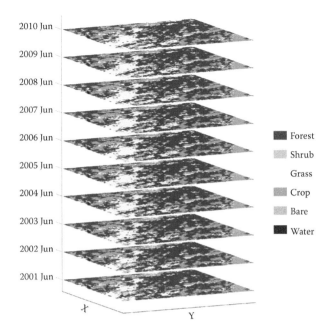

FIGURE 7.5
(**See color insert.**) Land cover maps in June of each year from 2001 to 2010.

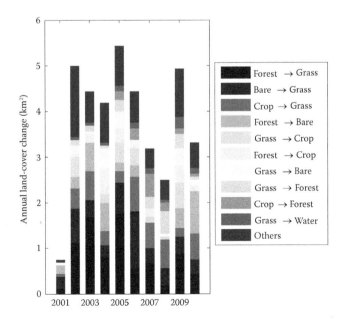

FIGURE 7.6
(**See color insert.**) Histogram of the area (km²) of the ten most frequent annual land cover changes between 2001 and 2010.

TABLE 7.2

Accuracy of Classification Using Standard SVM
with Radial Basis Function (RBF) Kernel, Standard
SVM with Polynomial (Poly) Kernel, and the
Modified SVM Method (mSVM)

	SVM		
	RBF	**Poly**	**mSVM**
Overall accuracy	0.766	0.732	**0.787**
Kappa coefficient	0.661	0.609	**0.696**
Producer's accuracy			
Forest	0.847	**0.850**	0.835
Shrub	0.028	0.028	**0.630**
Grass	0.682	0.613	**0.693**
Crop	0.810	0.777	**0.849**
Bare soil	**0.770**	0.637	0.754
User's accuracy			
Forest	0.741	0.710	**0.797**
Shrub	0.200	0.125	**0.529**
Grass	0.721	0.670	**0.757**
Crop	**0.807**	0.787	0.786
Bare soil	**0.896**	0.869	0.807

The highest value in each row is marked with a bold face.

the other two methods. For the producer's accuracy, the modified SVM generated the highest accuracy for three (shrub, grass, and crop) of the five classes. For the user's accuracy, it has the highest accuracy for the classes of forest, shrub, and grass. Especially for the shrub class, both producer's and user's accuracies for the modified SVM are much greater than those for the standard SVM with RBF kernel and the standard SVM with polynomial kernel.

7.4.3 Accuracy Assessment of Trajectory Mapping Results

Table 7.3 shows the annual and overall trajectory accuracies which evaluated the accuracy of class sequences over time. The modified SVM generated the greatest annual trajectory accuracy for all ten years, ranging from 0.820 to 0.8973 with an average of 0.863. These values were more accurate than the standard SVM with RBF kernel by 0.03–0.06, and greater than standard SVM with polynomial kernel by 0.06–0.09. For the overall trajectory accuracy, the modified SVM is greater by 0.08 and 0.13 than the standard SVM with RBF and polynomial kernels, respectively. These comparisons indicate the superiority of the modified SVM in trajectory mapping over standard SVM with commonly used kernels.

TABLE 7.3

Evaluation of Trajectory Mapping with Annual Trajectory Accuracy and Overall Trajectory Accuracy Using Standard SVM with Radial Basis Function (RBF) Kernel, Standard SVM with Polynomial (Poly) Kernel, and the Modified SVM Method (mSVM)

	Year	SVM		mSVM
		RBF	**Poly**	
Annual trajectory accuracy	2001	0.820	0.791	**0.885**
	2002	0.819	0.790	**0.868**
	2003	0.843	0.807	**0.873**
	2004	0.855	0.810	**0.885**
	2005	0.839	0.798	**0.868**
	2006	0.839	0.794	**0.873**
	2007	0.824	0.783	**0.865**
	2008	0.783	0.742	**0.820**
	2009	0.807	0.766	**0.845**
	2010	0.803	0.754	**0.845**
Overall trajectory accuracy	2001–2010	0.725	0.676	**0.804**

The highest value in each row is marked with a bold face.

7.4.4 Comparison of Adaptive Time Series with Full Length Time Series

Table 7.4 shows the accuracy of the classifications for generating annual maps using two types of input features: full length time series versus adaptive time series. The classifications with adaptive time series using standard SVM and the modified SVM method generated higher overall accuracies and Kappa coefficients than the full length time series. The improvement over the results using full length time series was 0.01 and 0.03 in overall accuracy, and 0.02 and 0.06 in Kappa coefficient, for the adaptive time series using standard SVM and modified SVM method, respectively.

Analyzing the overall accuracy calculated for the annual maps, we can see that the adaptive time series with the modified SVM method generated the highest accuracy for almost all years except for years 2003 and 2010. Compared to the accuracy for the annual maps generated using the full length time series, the improvement by using adaptive time series with the modified SVM ranged from 0.001 to 0.08 except for 2003. Even the accuracy for adaptive time series with standard SVM was higher than that for full length time series by 0.01–0.05 for the annual maps in six years. Although the full length time series generated the highest accuracy for the annual map in 2003, the difference is ignorable compared to the result for adaptive time series using the modified SVM method. In sum, the classifications using the adaptive time series generated more accurate results than using the full length time series for annual land cover mapping. In addition, the modified SVM

TABLE 7.4

Accuracy of Annual Land Cover Mapping Using Full Length Annual Time Series (Full) and Adaptive Time Series (Adaptive)

		Full	Adaptive	
		SVM	SVM	mSVM
Classification accuracy	Overall Accuracy	0.783	0.796	**0.816**
	Kappa	0.679	0.698	**0.732**
Accuracy calculated for each year	2001	0.820	0.768	**0.821**
	2002	0.770	0.803	**0.814**
	2003	**0.808**	0.780	0.800
	2004	0.835	0.843	**0.845**
	2005	0.786	0.779	**0.824**
	2006	0.757	0.811	**0.832**
	2007	0.790	0.787	**0.792**
	2008	0.730	0.767	**0.813**
	2009	0.787	0.828	**0.840**
	2010	0.745	**0.791**	0.776

The highest value in each row is marked with a bold face.

method performed better than the standard SVM method on classification with absent features.

7.5 Discussion

Traditional change detection studies are based on the analysis of two or multiple satellite images acquired on different dates with a relatively sparse temporal frequency (Lu et al. 2004). With dense time series of satellite imagery, it is possible to develop change detection techniques at higher temporal frequency as an approximation to continuous land cover monitoring. Indeed, as more frequent satellite observations become available, recent change detection studies have increasingly shifted from discrete snapshot approaches to continuous time series approaches (Lunetta et al. 2006; Kennedy et al. 2007; 2010; Huang et al. 2010; Jamali et al. 2015). Compared to traditional land cover mapping at yearly or decadal resolution, land cover mapping at higher temporal resolution, for instance, monthly, is much more difficult.

In this study, we proposed an integrated change detection and classification approach to map land cover trajectories at each pixel over the time period of 2001–2010 using 32-day MODIS time series. The use of adaptive time series generated more accurate classification results for each year by removing potential disturbance from land cover changes occurring in the middle of a year. Furthermore, the modified SVM method showed great potential in handling absent features in adaptive time series. The continuous trajectory mapping generated satisfactory results and accuracy for each year and across the ten years, which also demonstrated the potential of satellite image time series such as MODIS data for monitoring land cover dynamics at high temporal resolution.

The integrated training and classification approach used in this study offered two advantages compared with traditional classification. First, the integrated approach required less computational time, in that only one training process was required for all years no matter how many years are involved. In contrast, traditional annual classification requires the classifier trained for each individual year and the costs increase with the number of years considered. Second, the integrated approach offered more reference samples for the training process by pooling together the reference samples acquired for each year. Those two advantages may benefit change detection using multi-temporal images, where training samples collected in one year may be used in the classifications for other years, and fewer classifications are needed to generate time series of land cover maps.

Meanwhile, the effectiveness of the integrated approach also relied on both the quality of image data and the classifier. In this study, the MODIS NBAR product used as the input features are nadir BRDF adjusted reflectance with atmospheric correction, and consequently, radiometric difference in time series acquired from different years is expected to be very small or ignorable. The EVI calculated from the NBAR product is also insensitive to radiometric difference. Additionally, SVM classification method is only sensitive to support vectors on class boundaries in feature space and thus it is expected to be more robust to the within-class variations than other methods. Nevertheless, radiometric normalization would be crucial for the integrated approach using image data with evident radiometric difference even when they were acquired in the same seasons but in different years.

Both trajectory mapping and annual mapping were performed with the integrated classification approach and the modified SVM method to handle adaptive time series with absent features. Distinctively, the pixels with detected changes correspond to the longest adaptive time series in a year for annual mapping whereas for trajectory mapping, they were associated with more than one adaptive time series. Therefore, the classification using only the longest adaptive time series (for annual mapping) was expected to be more accurate than the classification using all available adaptive time series (for trajectory mapping) given that additional short time series may contribute to errors due to more absent features. This was consistent with the results in Table 7.2 (0.787 using all adaptive time series) and Table 7.4 (0.816 using only longest adaptive time series). On the other hand, the annual trajectory accuracy (Table 7.2) generated by trajectory mapping is greater than the annual accuracy (Table 7.4) generated by annual mapping, which was likely attributed to the temporal smoothing implemented after classification for trajectory mapping, where the temporal smoothing performed for each pixel within stable periods was able to correct some misclassification errors due to limited records in short adaptive time series. Similarly, comparison of Table 7.2 and Table 7.3 showed the overall trajectory accuracy (0.804) and annual trajectory accuracy for each year (0.82–0.88) were all higher than the classification accuracy (0.787) and much improved from the overall trajectory accuracy (0.38, unreported results) obtained before temporal smoothing. Therefore, it is evident that the temporal smoothing based on majority vote for a stable period is effective. Nevertheless, more advanced smoothing methods among neighboring time points could be investigated to further improve trajectory accuracy in future research.

Errors in land cover trajectory maps could arise from the results in both change date detection and classification of adaptive time series. This study only assessed the impact of classification errors by directly using the correct change dates derived from the reference pixels. To better understand the accuracy of trajectory mapping, a systematic accuracy evaluation may be developed to allow the combined evaluation of the change detection and classification in future study. Moreover, while temporal smoothing can reduce errors from both sources to some extent, spatial information may be explored in future study to further reduce errors in trajectory maps.

7.6 Conclusions

This study has demonstrated the proposed continuous trajectory mapping viable for monitoring land cover changes using dense time series such as MODIS data. The accuracy of trajectories for the study area was satisfactory across the ten years from 2001 to 2010 (generally greater than 0.84 with an average of 0.863). Additionally, identification of the land cover changes within a year led to greater accuracy for annual land cover mapping using adaptive time series than using full length time series. This study has shown the wealth of information in MODIS data and illustrated the potential of dense time series data for monitoring land cover at high temporal resolution.

This study provides a general framework for land cover trajectory mapping at high temporal resolution. The results from this study represent a baseline that can be achieved using a time series of 32-day average MODIS reflectance and EVI data. The use of shorter, 16-day or 8-day composite periods in MODIS data may be explored for the discrimination of subtle spectral-temporal differences among more land cover classes (e.g., different forest types). The land cover trajectory maps produced in this study represent a valuable source of information for many applications. Maps documenting land cover monthly can be used by scientists and policy makers to better understand the role of land use practices in various environmental issues. It is also hoped that the study will stimulate the use of MODIS data and similar data sources for land cover characterization at higher temporal resolution.

References

Alves, D.S., & Skole, D.L. 1996. Characterizing land cover dynamics using multi-temporal imagery. *International Journal of Remote Sensing*, 17, 835–839.

Cai, S., & Liu, D. 2015. Detecting change dates from dense satellite time series using a sub-annual change detection algorithm. *Remote Sensing*, 7, 8705–8727.

Chechik, G., Heitz, G., Elidan, G., Abbeel, P., & Koller, D. 2008. Max-margin classification of data with absent features. *Journal of Machine Learning Research*, 9, 1–21.

Coppin, P., Jonckheere, I., Nackaerts, K., Muys, B., & Lambin, E. 2004. Digital change detection methods in ecosystem monitoring: a review. *International Journal of Remote Sensing*, 25, 1565–1596.

Defries, R.S., & Belward, A.S. 2000. Global and regional land cover characterization from satellite data: An introduction to the Special Issue. *International Journal of Remote Sensing*, 21, 1083–1092.

Friedl, M. A., McIver, D. K., Hodges, J. C. F., Zhang, X. Y., Muchoney, D., Strahler, A. H., Woodcock, C. E., Gopal, S., Schneider, A., Cooper, A. et al. 2002. Global land cover mapping from MODIS: Algorithms and early results. *Remote Sensing of Environment* 83(1-2): 287–302.

Friedl, M. A., Sulla-Menashe, D., Tan, B., Schneider, A., Ramankutty, N., Sibley, A., & Huang, X. M. 2010. MODIS Collection 5 global land cover: Algorithm refinements and characterization of new datasets. *Remote Sensing of Environment* 114(1): 168–182.

Foley, J. A., DeFries, R., Asner, G. P., Barford, C., Bonan, G., Carpenter, S. R., Chapin, F. S., Coe, M. T., Daily, G. C., Gibbs, H. K. et al. 2005. Global consequences of land use. *Science* 309(5734): 570–574.

Gerten, D., S. Schaphoff, U. Haberlandt, W. Lucht & S. Sitch. 2004. Terrestrial vegetation and water balance—hydrological evaluation of a dynamic global vegetation model. *Journal of Hydrology* 286(1–4): 249–270.

Huang, C., Goward, S.N., Masek, J.G., Thomas, N., Zhu, Z., & Vogelmann, J.E. 2010. An automated approach for reconstructing recent forest disturbance history using dense Landsat time series stacks. *Remote Sensing of Environment*, 114, 183–198.

Jamali, S., Jonsson, P., Eklundh, L., Ardo, J., & Seaquist, J. 2015. Detecting changes in vegetation trends using time series segmentation. *Remote Sensing of Environment*, 156, 182–195.

Justice, C.O., Townshend, J.R.G., Vermote, E.F., Masuoka, E., Wolfe, R.E., Saleous, N., Roy, D.P., & Morisette, J.T. 2002. An overview of MODIS Land data processing and product status. *Remote Sensing of Environment*, 83, 3–15.

Kennedy, R.E., Cohen, W.B., & Schroeder, T.A. 2007. Trajectory-based change detection for automated characterization of forest disturbance dynamics. *Remote Sensing of Environment*, 110, 370–386.

Kennedy, R.E., Yang, Z., & Cohen, W.B. 2010. Detecting trends in forest disturbance and recovery using yearly Landsat time series: 1. LandTrendr -- Temporal segmentation algorithms. *Remote Sensing of Environment*, 114, 2897–2910.

Lambin, E., Baulies, X., Bockstael, N., Fischer, G., Krug, T., Leemans, R., Moran, E., Rindfuss, R., Sato, Y., & Skole, D. 1999. Land-use and land-cover change (LUCC): Implementation strategy. IGBP Report No. 48 and IHDP Report No. 10. C. Nunes and J. I. Augé.

Liu, D., & Cai, S. 2012. A spatial-temporal modeling approach to reconstructing land cover change trajectories from multi-temporal satellite imagery. *Annals of the Association of American Geographers*, 102, 1329–1347.

Lu, D., Mausel, P., Brondizio, E., & Moran, E. 2004. Change detection techniques. *International Journal of Remote Sensing*, 25, 2365–2407.

Lunetta, R.S., Knight, J.F., Ediriwickrema, J., Lyon, J.G., & Worthy, L.D. 2006. Land cover change detection using multi-temporal MODIS NDVI data. *Remote Sensing of Environment*, 105, 142–154.

Meyer, W. B., & Turner, I., B.L. (eds) 1994. *Changes in Land Use and Land Cover: A Global Perspective.* Cambridge, Cambridge University Press.

MODIS 1999. MODIS Vegetation Index (MOD 13): Algorithm Theoretical Basis Document (version 3) (http://modis.gsfc.nasa.gov/data/atbd/atbd_mod13.pdf). In.

Myhre, G., M. M. Kvalevag & C. B. Schaaf. 2005. Radiative forcing due to anthropogenic vegetation change based on MODIS surface albedo data. *Geophysical Research Letters* 32(21): L21410.

Rindfuss, R. R., S. J. Walsh, B. L. Turner, J. Fox & V. Mishra. 2004. Developing a science of land change: Challenges and methodological issues. *Proceedings of the National Academy of Sciences of the United States of America* 101(39), 13976–13981.

Salberg, A.B., & Jenssen, R. 2012. Land cover classification of partly missing data using support vector machines. *International Journal of Remote Sensing*, 33, 4471–4481.

Schaaf, C.B., Gao, F., Strahler, A.H., Lucht, W., Li, X.W., Tsang, T., Strugnell, N.C., Zhang, X.Y., Jin, Y.F., Muller, J.P. et al. 2002. First operational BRDF, albedo nadir reflectance products from MODIS. *Remote Sensing of Environment*, 83, 135–148.

Singh, N., & Bhaduri, B. L. 2010. Estimating the effect of biofuel on land cover change using multi-year MODIS land cover data. *2010 IEEE International Geoscience and Remote Sensing Symposium*, 899–902.

Steffen, W., Sanderson, A., Tyson, P. D., Jager, J., Matson, P. A., Moore, B. I., Oldfield, F., Richardson, K., Schellnhuber, H. J., Turner, B. L. I. et al. 2004. *Global Change and the Earth System: A Planet under Pressure.* Berlin, Springer.

Turner, B. L. I., Clark, W. C., Kates, R. W., Richards, J. F., Mathews, J. T., & Meyer, W. B. (eds) 1990. *The earth as Transformed by Human Action: Global and Regional Changes in the Biosphere Over the Past 300 Years.* Cambridge, Cambridge Univ. Press.

Turner, W. R., Brandon, K., Brooks, T. M., Gascon, C., Gibbs, H. K., Lawrence, K. S., Mittermeier, R. A., & Selig, E. R. 2012. Global Biodiversity Conservation and the Alleviation of Poverty. *Bioscience* 62(1): 85–92.

Verbesselt, J., Hyndman, R., Newnham, G., & Culvenor, D. 2010. Detecting trend and seasonal changes in satellite image time series. *Remote Sensing of Environment*, 114, 106–115.

Vitousek, P. M., Mooney, H. A., Lubchenco, J., & Melillo, J. M. 1997. Human domination of Earth's ecosystems. *Science* 277(5325): 494–499.

Wiedinmyer, C., Akagi, S. K., Yokelson, R. J., Emmons, L. K., Al-Saadi, J. A., Orlando, J. J., & Soja, A. J. 2011. The Fire INventory from NCAR (FINN): a high resolution global model to estimate the emissions from open burning. *Geoscientific Model Development* 4(3): 625–641.

Zhou, F., Zhang, A., & Townley-Smith, L. 2013. A data mining approach for evaluation of optimal time-series of MODIS data for land cover mapping at a regional level. *Isprs Journal of Photogrammetry and Remote Sensing*, 84, 114–129.

Zhu, Z., & Woodcock, C.E. 2014. Continuous change detection and classification of land cover using all available Landsat data. *Remote Sensing of Environment*, 144, 152–171.

8

Creating a Robust Reference Dataset for Large Area Time Series Disturbance Classification

Mariela Soto-Berelov, Andrew Haywood, Simon Jones, Samuel Hislop, and Trung H. Nguyen

CONTENTS

8.1 Introduction 157
8.2 Study Area 158
8.3 Methods 159
    8.3.1 Quality Control and Quality Assurance 161
8.4 Case Study 162
    8.4.1 Quality Control and Quality Assurance 164
    8.4.2 Mapping Disturbance 165
8.5 Discussion and Conclusion 168
Acknowledgments 169
References 170

8.1 Introduction

Change detection methods have evolved from mapping and identifying change between image pairs, to sophisticated techniques based on dense time series image datasets. This paradigm shift of bi-temporal change detection to multi-temporal time series analysis has been accompanied by a surge in algorithms capable of extracting the spectral trajectories of pixels over time (Huang et al. 2010; Kennedy, Yang, and Cohen 2010; Verbesselt et al. 2010; Zhu and Woodcock 2014; Schmidt et al. 2015). These techniques capture spectral change signals associated with processes occurring on the ground, after filtering out seasonal or year to year noise introduced by many factors, including sun angle, sensor drift, haze, and phenology. Change metrics that can be detected within a particular time period include date of onset, direction (positive or negative), magnitude, duration, and recovery information. Although such metrics are extremely useful, information regarding the change agent needs to be acquired separately through a reference dataset. For example, if interested in changes caused by fire, a dataset containing the location of fire sightings can be used for classification (Schroeder et al. 2011).

Reference datasets can be used for both training and validating models. They can be used as classifiers (e.g., decision trees, maximum likelihood) to fit and evaluate a model (Broich et al. 2011; Haywood, Verbesselt, and Baker 2016). These can be based on existing datasets (e.g., fire databases, forest inventory plots) or a combination of ancillary sources (e.g., Google Earth, Rapid Eye, corporate datasets, field plots) that are assembled using a human interpreter approach (Cohen, Yang, and Kennedy 2010; DeVries et al. 2015). One concern with the use of reference data is that it is often incomplete and biased. Aware of this restriction, some authors have developed data collection tools such as TimeSync which ensure samples are statistically robust and free of temporal and spatial restrictions (Cohen et al. 2010). While such methods can produce reliable results, they are generally more suitable for small area applications.

This chapter presents good practice guidelines for the creation of a reference dataset that takes advantage of an existing forest inventory plot network. The reference dataset consists of 7860 reference pixels over a large area containing public land forests in the state of Victoria, Australia. The reference dataset is built around an extensive forest inventory plot network (Haywood, Mellor, and Stone 2016) stratified by biogeographic regions and public land tenure (e.g., National Park and State Forest). The advantage of using this approach is that it ensures the dataset is unbiased to a particular location and is comprehensive over a large area. In this case study, the method applied captures all major geographical regions in Victoria. A combination of different types of ancillary datasets ranging from Google Earth imagery, the pixel's trajectory as extracted from annual composites, local expert knowledge, and regional spatial datasets are used by trained interpreters to attribute meaningful disturbance information. To demonstrate the utility of the reference dataset, a subset is then used in a machine learning environment to produce classified disturbance maps over a 28 year period according to agent and severity categories.

8.2 Study Area

The study area, shown in Figure 8.1, comprises all forests located on public land across the state of Victoria, Australia (over 8.1 million hectares). These are spread across 11 biogeographic regions (Interim Biogeographic Regionalisation for Australia or IBRA) (Cummings and Hardy 2000), which differ in elevation, temperature, and rainfall. To monitor and report across a range of forest indicators, the state of Victoria's land management agency

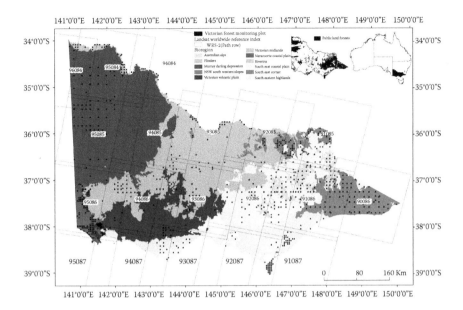

FIGURE 8.1
(**See color insert.**) Map of Victoria showing bioregions and the Victorian Forest Monitoring Plot (VFMP) Network across the public forest estate.

has established a state wide forest inventory network using a fixed sampling design (Victorian Forest Monitoring Program or VFMP) (Haywood, Mellor, and Stone 2016). It contains 786 forest inventory plots, which were established by creating a grid across the state, and randomly selecting plots stratified according to biogeographic region and public land tenure (multiple use forests consisting mainly of state forests and nature conservation reserves that mostly contain national parks). The inventory network is also based on multi-staged sampling that follows a three tiered system. At the highest level, it includes ground plot data periodically collected within a 15 m radius within each plot. At the intermediate level, a 2 km × 2 km area surrounding each ground plot has been photo interpreted using high resolution aerial imagery (Farmer et al. 2011). At the lowest level, Landsat and MODIS imagery have been used to map forest extent, using training data from each 2 km × 2 km interpreted photo-plot (Mellor et al. 2013).

8.3 Methods

To attribute spectral change into meaningful disturbance agents such as logging, fire, and disease, a reference dataset consisting of 7860 reference pixels was created. These were randomly selected pixels built around the aforementioned VFMP that contains 786 plots scattered across the state (Farmer et al. 2011; Haywood, Mellor, and Stone 2016). Within each plot, 10 reference points (each representing a 30 m pixel) were randomly chosen with a minimum distance of 250 m between them. To avoid misregistration errors, we established a 3 × 3 window surrounding each pixel of interest. If a reference pixel fell on the boundary of two land cover types (e.g., shrubland and forest) or at the edge of a disturbance, the reference pixel was moved away from the edge.

The spectral change signal of forest disturbance can usually be interpreted by an experienced image analyst (Masek et al. 2008; Huang et al. 2010), especially when having access to images that precede and succeed the disturbance. Each reference pixel was assessed by a trained interpreter using a Multiple Lines of Evidence (MLE) approach (Figure 8.2) to determine whether the pixel had been disturbed during the 1988–2016 time period. Several sources of information were used, in addition to Landsat data, to create the reference dataset (Table 8.1). Among others, these include Google Earth Imagery, corporate spatial datasets such as the State Fire History

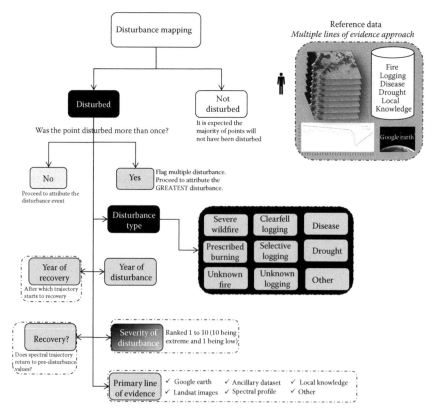

FIGURE 8.2
Disturbance attribution obtained by interpreter using a multiple lines of evidence approach.

TABLE 8.1

Reference Datasets Used for the Attribution of Disturbance Events

Evidence	Description
Ancillary dataset—logging	Corporate dataset that includes the location of logging events.
Ancillary dataset—fire history	Corporate dataset that includes the location of fire events.
Spectral trajectory	Time series trajectory of TCW and NDVI spectral indices (1988–2015) on a per pixel basis.
Annual composites of TCW	Time series (1988–2015) rasters of a spectral index/ transformation obtained from the Landsat Thematic Mapper (TM) on a per pixel basis.
Google Earth imagery	The location of the point was examined in Google Earth with higher resolution imagery. Where historical imagery was available, it was also examined as an additional reference source to determine any disturbance events.
Landsat images	One cloud free image taken during summer was used per year, viewed as a false color composite.
Local knowledge	Information obtained from interviews with park rangers or local land managers. For example: die back events associated with koalas or insect infestation; rainfall graphs that highlight years of drought.

Database (Department of Environment and Primary Industries 2016a) and the State Logging History Database (Department of Environment and Primary Industries 2016b), and the per-pixel spectral trajectory of a spectral index and transformation commonly used in time series analyses given their responsiveness to vegetation disturbance (Normalized Difference Vegetation Index or NDVI and Tasseled Cap Wetness or TCW).

Given what is known of the disturbance history in Victoria, we expected to find disturbances in approximately one third of the reference pixels. We also expected to find hotspots of change in parts of the state. Whenever a disturbance event was identified by an interpreter, the type of disturbance, severity of disturbance, and year of disturbance was recorded in a database. The information collected for each reference pixel is shown in Figure 8.2.

8.3.1 Quality Control and Quality Assurance

To ensure the reference dataset is of high quality several measures were implemented. Interpreters were initially trained by working on reference pixels within a training area (these were only used for training). A standard operating procedure and reference dataset protocol were created and regularly updated to incorporate lessons learned during the duration of the reference dataset creation. In addition, interpreters were blindly assigned overlapping pixels (every tenth) for quality control purposes. Overlapping pixels were assessed for consistency and findings were discussed with the team of interpreters during weekly meetings. During meetings, interpreters also discussed pixels that were difficult to interpret or were good examples of different disturbance types. These measures enhanced the quality of the attribution, which saw improvement throughout the duration of the project. To avoid a geographic bias on quality, pixels were assigned across the state instead of by region.

In addition to quality control, an independent experienced interpreter assessed pixels for quality assurance. This was done by evaluating the accuracy of the reference dataset by randomly interpreting 10% of all reference pixels. Overall accuracy, user's, and producer's accuracy were examined in this way.

8.4 Case Study

A subset of the reference dataset located in the eastern part of the state was used to extract disturbance trends over the last 28 years and create classified maps of disturbance (Figure 8.3).

The area falls within the Worldwide Reference System (WRS) tile Path 90 Row 86. It is located within the South East Corner bioregion, which contains rugged topography traversed by rivers that empty into Bass Strait. The area is densely forested, containing a mixture of forest types, including high elevation woodlands, wet and damp sclerophyll forests intermixed with arid woodlands, lowland and coastal sclerophyll forests, and warm temperate rainforests. This

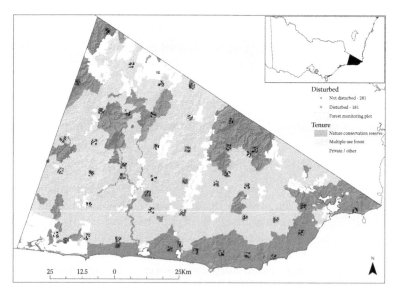

FIGURE 8.3
Location of case study area in Eastern Victoria.

TABLE 8.2

Land Tenure within Case Study Area

Tenure	Area (ha)
Multiple use forest	424,916
Nature conservation reserve	197,294
Other public land	8694
Private/other (non-public)	71,362

area was selected since it mainly consists of publically owned land (Table 8.2) and is densely forested. While the majority of the area contains multiple use forests, there are several national, regional, and state parks (all of which fall under the nature conservation reserve category). In addition, this area has experienced dynamic changes resulting from the disturbance agents most common across Victorian forests—logging and fire (both wildfires and planned). Within the study area, logging activity is common practice across multiple use forests, while planned burning is carried out throughout all public land.

A total of 47 inventory plots and 462 reference pixels (Figure 8.3) fall within the study area (8 reference pixels fell within the adjacent WRS tile and do not form part of the study area). Using the MLE approach described earlier, interpreters virtually assessed these pixels and found that 39% were disturbed at some point between 1988 and 2016. Table 8.3 lists disturbance according to agent and severity, as attributed by interpreters.

The majority of the disturbed reference pixels were impacted by wildfires or prescribed burns. Prescribed burns are controlled fires used to reduce fuel loads and lessen the potential catastrophic effect of wildfires. Logging activities (mainly clear fell logging) were also recorded in 13% of the reference pixels within state forests. Less than 5% of the reference pixels were classified as having experienced low severity disturbance caused by drought or unknown factors.

TABLE 8.3

Attributed Reference Pixels

Type	Description	N	Severity Low	Medium	High
0	No disturbance	281			
1	Other (low severity)	17	17		
2	Selective logging	17	1	15	1
3	Clear fell logging	44		3	41
4	Prescribed burn	59	43	15	1
5	Wildfire	44	8	25	11
Total		462	69	58	54

Severity ranking is derived from interpreter attribution and consists of Low (1–3), Medium (4–7), and High (8–10).

8.4.1 Quality Control and Quality Assurance

As outlined earlier, a series of reference pixels were blindly interpreted by several operators for quality control. Table 8.4 shows there was 77% agreement between interpreters. Of the 33% disagreement, 13% was caused by reference pixels that were attributed as low severity disturbance (Table 8.5). When considering reference pixels that were attributed with higher severity disturbance (severity > 2), agreement significantly improved (90%).

Quality assurance provided an additional error assessment. This was completed by an external expert interpreter, who evaluated 10% of the reference pixels. Tables 8.6 and 8.7 show interpreter accuracy when considering all severity categories (Table 8.6) and excluding those disturbances assigned a low severity (Table 8.7).

The reference dataset obtained an overall accuracy of 80% when considering all disturbed reference pixels, regardless of severity assigned. Accuracy was much greater (Table 8.7) if only considering as disturbed those reference pixels that have a severity greater than 2. Results from quality control and quality assurance suggest that disturbed pixels are correctly classified over 90% of the time when they have a severity higher than 3. Low severity disturbance events, on the other hand, had less consistency between interpreters.

TABLE 8.4

Error Matrix Showing Accuracy Obtained by Interpreters When Attributing the Same Reference Pixels across All Severity Categories

Disturbed Category Includes All Severities		Interpreter				
		Disturbed	Not Disturbed	Row Total	Accuracy	%
QC	Disturbed	19	11	30	Overall accuracy	76.7
	Not disturbed	3	27	30		
	Column total	22	38	60		

TABLE 8.5

Error Matrix Showing Accuracy Obtained by Interpreters When Attributing the Same Reference Pixels, but Excluding Those Attributed with a Disturbance Severity Smaller Than 3

Disturbed Category Does not Include Disturbance Events of Low Severity (<3)		Interpreter				
		Disturbed	Not Disturbed	Row Total	Accuracy	%
QC	Disturbed	19	4	23	Overall accuracy	90.0
	Not disturbed	2	35	37		
	Column total	21	39	60		

TABLE 8.6

Error Matrix Showing Accuracy Obtained by Interpreters When Attributing
Reference Pixels across All Severity Categories

Disturbed Category Includes All Severities		Interpreter				
		Disturbed	Not Disturbed	Row Total	Accuracy	%
QA	Disturbed	17	8	25	Overall accuracy	80.4
	Not disturbed	1	20	21	Omission errors	32.0
	Column total	18	28	46	Commission errors	5.6

TABLE 8.7

Error Matrix Showing Accuracy Obtained by Interpreters

Disturbed Category Does not Include Disturbance Events of Low Severity (<3)		Interpreter				
		Disturbed	Not Disturbed	Row Total	Accuracy	%
QA	Disturbed	17	1	18	Overall accuracy	97.8
	Not disturbed	0	28	28	Omission errors	5.6
	Col total	17	29	46	Commission errors	0.0

Disturbed pixels exclude those with a disturbance severity smaller than 3.

8.4.2 Mapping Disturbance

The reference pixels were used in a Random Forest classification (Breiman 2001) implemented through the R statistics package (R Development Core Team 2011), to create attributed disturbance maps during a 28 year period (1988–2016). First, images from the Landsat Thematic Mapper (TM) and Enhanced Thematic Mapper (ETM+) sensors taken between December and March (southern hemisphere summer) for Landsat WRS2 Tile Path 90 Row 86 were obtained from the USGS archive as Level-1 terrain corrected (L1T) surface reflectance products, derived via The Landsat Ecosystem Disturbance Adaptive Processing System or LEDAPS (Masek et al. 2013). Using the LandsatLinkr package (Braaten et al. 2015), a per-pixel annual time series was built by creating summer composites of Tasseled Cap Wetness (TCW) and the Normalized Burn Ratio index (NBR). The annual composites were then used to extract spectral change maps using the LandTrendr (Landsat-based Detection of Trends in Disturbance and Recovery) algorithm (Kennedy, Yang, and Cohen 2010). LandTrendr fits straight line segments to the temporal trajectory of each pixel. This allows the spectral trajectory to be characterized over time as having undergone no change, positive, and/or negative change.

To create disturbance maps, a series of predictor variables, consisting of elements derived from the LandTrendr outputs, including magnitude, year (also used to subsequently map the year of disturbance), and duration of segments were used. In addition, topographic and climatic variables commonly used when modelling vegetation dynamics were also included (Duffy et al. 2007; Powell et al. 2010; Frazier et al. 2014). These include average rainfall, temperature, elevation, slope, and aspect.

The disturbance categories mapped are shown in Table 8.3. They consist of six categories: no disturbance, low severity disturbance caused by an agent other than logging or fire, selective logging, clear fell logging, prescribed burn, and wildfire. The logging and fire categories mapped were grouped according to agent, regardless of severity. For instance, prescribed burning consists of 59 reference pixels, most of which had an associated low severity assigned by interpreters. Clear fell logging on the other hand consists mostly of reference pixels that were assigned a high severity of disturbance. This classification scheme was chosen to ensure classes had sufficient samples.

The RF model produced an overall accuracy of 68%, according to the out-of-bag error assessment. Omission rates varied amongst the categories from very high for the *Other low severity disturbance* and *Selective logging categories* (100%) to low for the *Clear fell logging* class (36%).

Given the observed difficulty of classifying disturbance events with severity rankings lower than 3, a second model was run that treated such reference pixels as undisturbed. This second model produced an overall accuracy of 73% (the error matrix is shown in Table 8.8).

The two classes that obtained the highest omission errors are *Other low severity disturbance* and *Prescribed burn*. Both these classes had most reference pixels assigned severity attributions that were less than 5 (10 being most severe). Unsurprisingly, the class that contained the highest number of reference pixels attributed as high severity disturbance (clear fell logging) obtained the lowest omission errors. The classified disturbance map according to agent and year of disturbance is shown in Figure 8.4.

As shown in Figure 8.4 (Top), areas impacted by wildfire concentrate along the northwestern part of the study area. In agreement with official state records, these mostly result from a 2014 wildfire, although some areas correspond with 2003 and 2004 wildfires. Planned burning was mostly detected across

TABLE 8.8

Error Matrix Produced by the Random Forests Classification, Removing as Disturbed Those with Severity Lower Than 3

	0	1	2	3	4	5	Omission Error
0	311	0	1	5	1	4	0.03
1	10	0	0	1	0	0	1
2	9	0	2	6	0	0	0.88
3	10	0	4	29	0	1	0.34
4	25	0	0	2	0	0	1
5	24	0	0	3	0	14	0.65

Classification refers to: 0—No disturbance; 1—Other low severity disturbance; 2—Selective logging; 3—Clear fell logging; 4—Prescribed burn; 5—Wildfire.

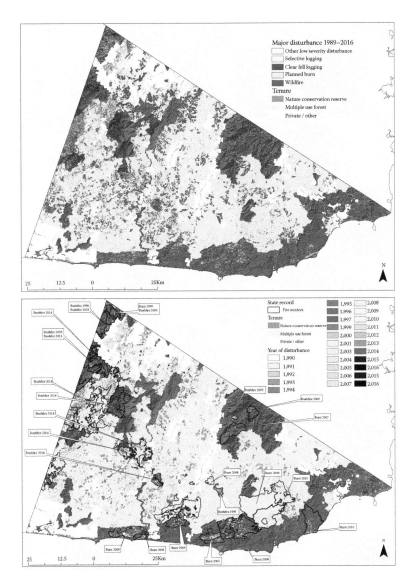

FIGURE 8.4
(**See color insert.**) Pixels that experienced a major disturbance during 1988 to 2016 (Top) and year of disturbance (Bottom). Labels in bottom refer to fire incidents (fire type and year) that are present in the state records.

the southern portion of the study area in both multiple use forests and nature conservation reserves during the years 2005–2010. Although omission rates for planned burns were extremely high (most reference pixels that fell within this category had an associated low severity), predicted areas tend to agree with state records (Figure 8.4 Bottom).

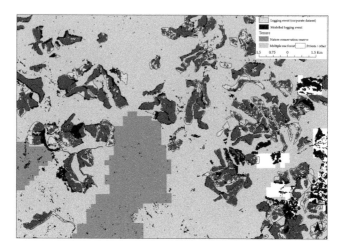

FIGURE 8.5
Areas predicted to have been disturbed by logging (black) and areas that according to the state records were logged.

Logging, which obtained the lowest omission rates, was detected across the study area (Figure 8.4 Top), mostly within multiple use forest areas (State Forests). As shown in Figure 8.5, overlap with state logging records was high. The year predicted by the model also closely matched that in the state logging records.

Figure 8.6 shows area predicted to have been logged from 1990 to 2016, according to land tenure. This figure confirms that most logging occurs within multiple use forests, followed by logging within privately owned lands. In addition, it indicates years of increased logging activity, while it also shows trends over time. For instance, logging in multiple use forests has been declining over the last 10 years.

8.5 Discussion and Conclusion

Reference data is essential for interpreting spectral time series and accurately characterizing large area disturbance and recovery events. Successfully attributing disturbance events requires a significant amount of human input. Logistically, it can be expensive to produce high quality reference datasets and often the question arises as to what constitutes an appropriate reference dataset for the analysis extent and classes that are being attributed. Often a reference dataset is chosen depending on data availability. The limitation of this approach is that the reference dataset can be geographically and temporally biased.

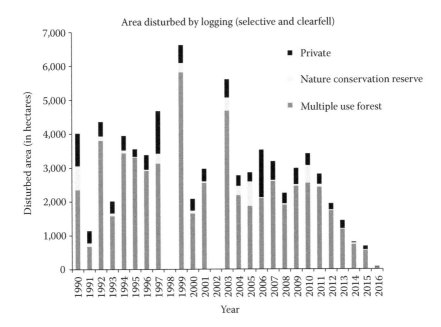

FIGURE 8.6
Total logged area during 1990–2016 according to land tenure.

In this chapter, we have presented a framework for creating a robust reference dataset that takes advantage of an existing large area sample-based strategic forest inventory network. This method is recommended when the objective is to derive large area disturbance mapping where a forest inventory network is already in place. The reference dataset, therefore, inherits the systematic sampling framework and consistency of the inventory network. We have shown here that human interpreters, using a multiple lines of evidence approach, can create a robust and comprehensive disturbance database according to agent and severity. However, low severity events remain challenging to both detect and accurately classify.

Building a reference dataset in this way can support large area time series mapping, which in turn complements forest inventory demands and allows land managers to assess and report on the condition of the forest estate in a holistic and consistent way.

Acknowledgments

We would like to thank the reviewers for their valuable comments, which helped improve the manuscript. We also acknowledge the work of trained interpreters

Salahuddin Ahmad, Shirley Famelli, Jenna Guffogg, Eloise Thaisen, and Daisy San Martin Saldias. This research was funded by the Cooperative Research Centre for Spatial Information (CRCSI) under Project 4.104 (A Monitoring and Forecasting Framework for the Sustainable Management of SE Australian Forests at the Large Area Scale). CRCSI activities are funded by the Australian Commonwealth's Cooperative Research Centres Programme.

References

Braaten, J, W Cohen, Z Yang, and E Haunreiter. 2015. *"LandsatLinkr 0.1.6 User Guide."* 1–41.

Breiman, L. 2001. "Random Forest." *Machine Learning* 45, 5–32. doi: 10.1023/A: 1010933404324.

Broich, M, MC Hansen, P Potapov, B Adusei, E Lindquist, and SV Stehman. 2011. "Time-Series Analysis of Multi-Resolution Optical Imagery for Quantifying Forest Cover Loss in Sumatra and Kalimantan, Indonesia." *International Journal of Applied Earth Observation and Geoinformation* 13 (2): 277–91. Elsevier B.V. doi: 10.1016/j.jag.2010.11.004

Cohen, WB, Z Yang, and R Kennedy. 2010. "Detecting Trends in Forest Disturbance and Recovery Using Yearly Landsat Time Series: 2. TimeSync - Tools for Calibration and Validation." *Remote Sensing of Environment* 114 (12): 2911–24. Elsevier B.V. doi: 10.1016/j.rse.2010.07.010

Cummings, B, and A Hardy. 2000. *"Revision of the Interim Biogeographic Regionalisation for Australia (IBRA) and Development of Version 5.1."* Canberra.

Department of Environment and Primary Industries. 2016a. *"Fire History Records of Fires Primarily on Public Land (FIRE_HISTORY)."* Department of Environment and Primery Industries.

Department of Environment and Primary Industries. 2016b. *"Harvested Logging Coupes (LOG_SEASON)."* Department of Environment and Primary Industries.

DeVries, B, M Decuyper, J Verbesselt, A Zeileis, M Herold, and S Joseph. 2015. "Tracking Disturbance-Regrowth Dynamics in Tropical Forests Using Structural Change Detection and Landsat Time Series." *Remote Sensing of Environment* 169: 320–34. Elsevier Inc. doi: 10.1016/j.rse.2015.08.020

Duffy, PA, J Epting, JM Graham, TS Rupp, and AD McGuire. 2007. "Analysis of Alaskan Burn Severity Patterns Using Remotely Sensed Data." *International Journal of Wildland Fire.* 16 (3), 277–84. doi: 10.1071/WF06034.

Farmer, E, S Jones, M Clarke, M Soto-Berelov, A Mellor, and A Haywood. 2011. "Semi-Automated API for Large Area Public Land Monitoring and Reporting." In *Proceedings of the GSR_1 Research Symposium*, Melbourne, Australia, 12–14 December 2011, 1–12.

Frazier, RJ, NC Coops, MA Wulder, and R Kennedy. 2014. "Characterization of Aboveground Biomass in an Unmanaged Boreal Forest Using Landsat Temporal Segmentation Metrics." *ISPRS Journal of Photogrammetry and Remote Sensing* 92: 137–46. International Society for Photogrammetry and Remote Sensing, Inc. (ISPRS). doi: 10.1016/j.isprsjprs.2014.03.003

Haywood, A, A Mellor, and C Stone. 2016. "A Strategic Forest Inventory for Public Land in Victoria, Australia." *Forest Ecology and Management* 367 (May): 86–96. doi: 10.1016/j.foreco.2016.02.026

Haywood, A, J Verbesselt, and PJ Baker. 2016. "Mapping Disturbance Dynamics in Wet Sclerophyll Forests Using Time Series Landsat." *ISPRS - International Archives of the Photogrammetry, Remote Sensing and Spatial Information Sciences* XLI-B8 (July): 12–9. doi: 10.5194/isprsarchives-XLI-B8-633-2016

Huang, C, SN Goward, JG Masek, N Thomas, Z Zhu, and JE Vogelmann. 2010. "An Automated Approach for Reconstructing Recent Forest Disturbance History Using Dense Landsat Time Series Stacks." *Remote Sensing of Environment* 114 (1): 183–98. Elsevier Inc. doi: 10.1016/j.rse.2009.08.017

Kennedy, RE, Z Yang, and WB Cohen. 2010. "Detecting Trends in Forest Disturbance and Recovery Using Yearly Landsat Time Series: 1. LandTrendr—Temporal Segmentation Algorithms." *Remote Sensing of Environment* 114 (12): 2897–910. Elsevier Inc. doi: 10.1016/j.rse.2010.07.008

Masek, JG, C Huang, R Wolfe, W Cohen, F Hall, J Kutler, and P Nelson. 2008. "North American Forest Disturbance Mapped from a Decadal Landsat Record." *Remote Sensing of Environment* 112: 2914–26. doi: 10.1016/j.rse.2008.02.010

Masek, JG, EF Vermote, N Saleous, R Wolfe, FG Hall, F Huemmrich, F Gao, J Kutler, and TK Lim. 2013. "*LEDAPS Calibration, Reflectance, Atmospheric Correction Preprocessing Code, Version 2.*" ORNL DAAC, Oak Ridge, Tennessee, USA. http://dx.doi.org/10.3334/ORNLDAAC/1146

Mellor, A, A Haywood, C Stone, and S Jones. 2013. "The Performance of Random Forests in an Operational Setting for Large Area Sclerophyll Forest Classification." *Remote Sensing* 5 (6): 2838–56. doi: 10.3390/rs5062838

Powell, SL, WB Cohen, SP Healey, RE Kennedy, GG Moisen, KB Pierce, and JL Ohmann. 2010. "Quantification of Live Aboveground Forest Biomass Dynamics with Landsat Time-Series and Field Inventory Data: A Comparison of Empirical Modeling Approaches." *Remote Sensing of Environment* 114 (5): 1053–68. Elsevier Inc. doi: 10.1016/j.rse.2009.12.018

R Core Team R: A Language and Environment for Statistical Computing. 2011.

Schmidt, M, R Lucas, P Bunting, J Verbesselt, and J Armston. 2015. "Multi-Resolution Time Series Imagery for Forest Disturbance and Regrowth Monitoring in Queensland, Australia." *Remote Sensing of Environment* 158: 156–68. Elsevier Inc. doi: 10.1016/j.rse.2014.11.015

Schroeder, TA, MA Wulder, SP Healey, and GG Moisen. 2011. "Mapping Wildfire and Clearcut Harvest Disturbances in Boreal Forests with Landsat Time Series Data." *Remote Sensing of Environment* 115 (6): 1421–33. Elsevier B.V. doi: 10.1016/j. rse.2011.01.022

Verbesselt, J, R Hyndman, G Newnham, and D Culvenor. 2010. "Detecting Trend and Seasonal Changes in Satellite Image Time Series." *Remote Sensing of Environment* 114 (1): 106–15. Elsevier B.V. doi: 10.1016/j.rse.2009.08.014

Zhu, Z, and CE Woodcock. 2014. "Continuous Change Detection and Classification of Land Cover Using All Available Landsat Data." *Remote Sensing of Environment* 144: 152–71. Elsevier Inc. doi: 10.1016/j.rse.2014.01.011

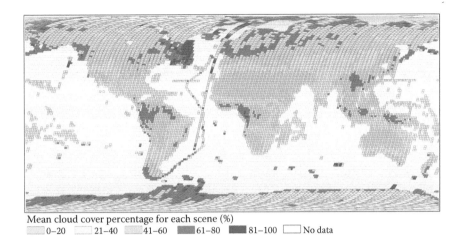

Mean cloud cover percentage for each scene (%)
☐ 0–20　☐ 21–40　☐ 41–60　■ 61–80　■ 81–100　☐ No data

FIGURE 1.1
Mean global cloud cover percentage calculated based on all available Landsat 8 images acquired between September 2013 and August 2017. A total of 966,708 Landsat 8 images are used. The mean global cloud cover percentage from all Landsat 8 observations is 41.59%.

FIGURE 2.3
False-color composite of 3 Landsat-7 images in Mona Island with low, medium, and high proportion of contaminated pixels (from left to right).

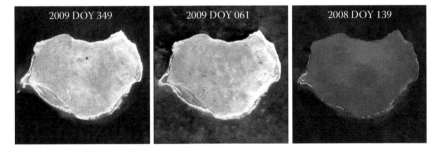

FIGURE 2.5
Reconstruction results of three contaminated images shown in Figure 2.3.

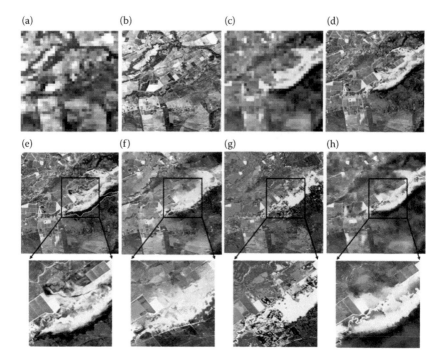

FIGURE 3.10

(a) MODIS image of November 26th; (b) Landsat image of November 26th; (c) MODIS image of December 12th; (d, e) Landsat image of December 12th; predicted images of December 12th by STARFM (f), UBDF (g), and FSDAF (h). White line in (e) delineates the boundaries of inundated area.

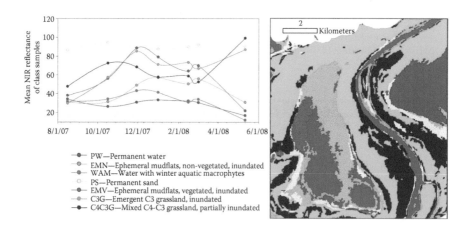

FIGURE 4.4

Example of seasonal trajectories of near infrared reflectance for a set of wetland dynamic cover types at Poyang Lake, China, (left) and their spatial representation in a portion of this wetland area (right).

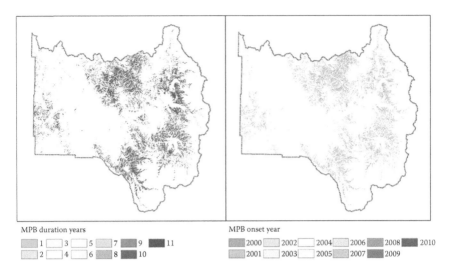

FIGURE 4.6
LandTrendr analysis results showing the duration (left) and onset year (right) of forest mortality attributed to mountain pine beetle infestation. (Based on the study by Liang, L. et al. 2014b. *Remote Sensing* 6, 5696–5716.)

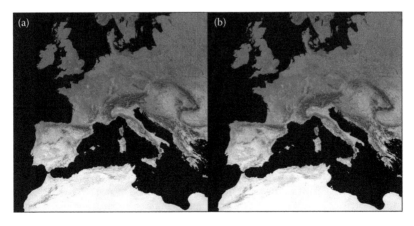

FIGURE 5.6
RGB composite of (a) collection-5 (YAST, MAST, Theta) and (b) collection-6 (YAST, MAST, Theta).

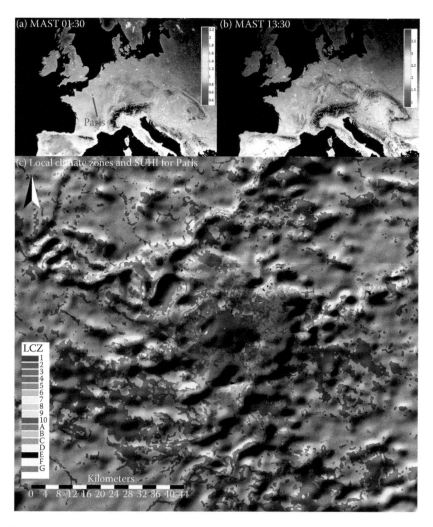

FIGURE 5.15
Mean annual SUHI at (a) night time and (b) daytime for Central Europe. (c) The nocturnal
annual mean SUHI of Paris overlaid as relief on a Local Climate Zone land cover classification
according to Stewart and Oke (2012). LCZ 2 = compact midrise (urban core); LCZ 6 = open low
rise (suburban housing); LCZ A = dense trees (forest).

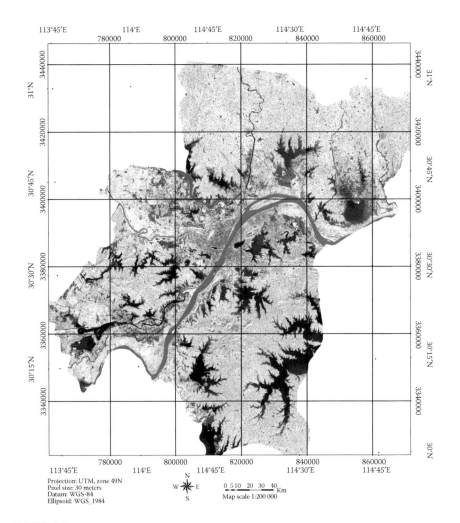

FIGURE 6.1
The geographic location of Wuhan city.

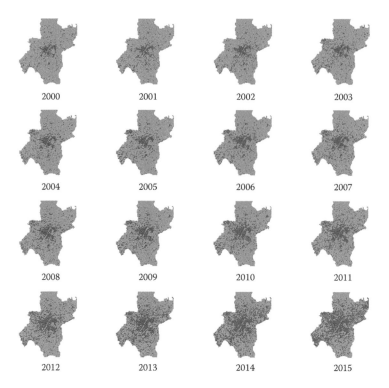

FIGURE 6.4
Annual maps of impervious surfaces in Wuhan city.

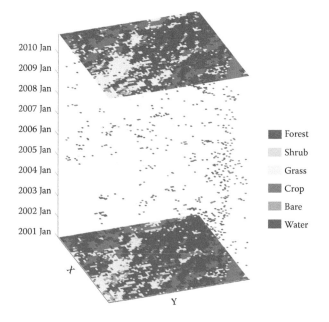

FIGURE 7.4
Land cover trajectories from 2001 to 2010. The plots at the bottom and the top correspond to the class maps for the starting point (January 2001) and ending point (December 2010), respectively. The land cover classes between the two time points are displayed only if land cover changes occurred.

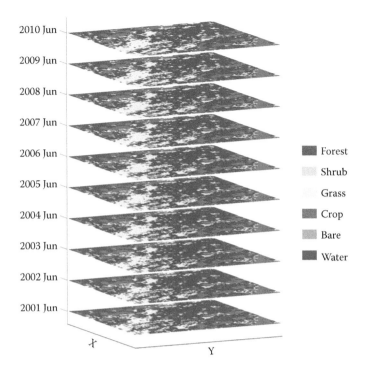

FIGURE 7.5
Land cover maps in June of each year from 2001 to 2010.

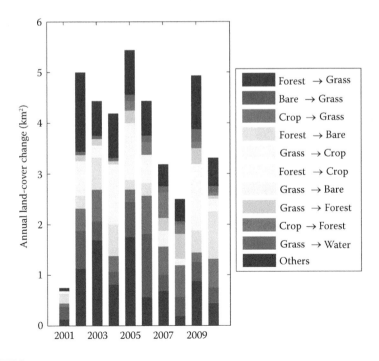

FIGURE 7.6
Histogram of the area (km²) of the ten most frequent annual land cover changes between 2001 and 2010.

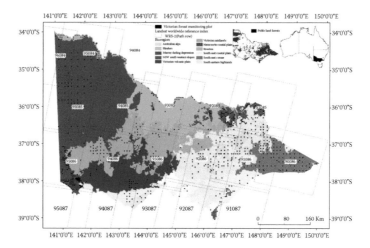

FIGURE 8.1
Map of Victoria showing bioregions and the Victorian Forest Monitoring Plot (VFMP) Network across the public forest estate.

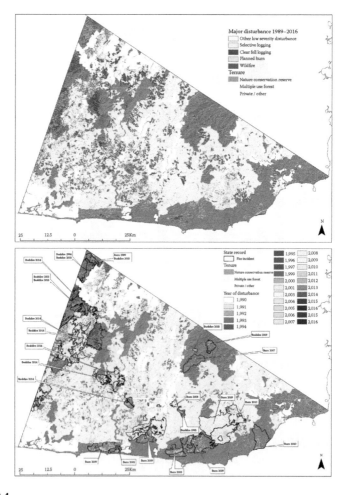

FIGURE 8.4
Pixels that experienced a major disturbance during 1988 to 2016 (Top) and year of disturbance (Bottom). Labels in bottom refer to fire incidents (fire type and year) that are present in the state records.

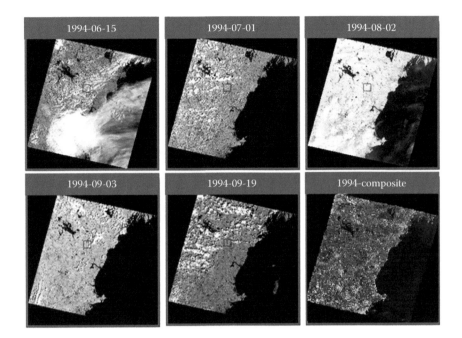

FIGURE 9.3
No cloud free image was available during the summer growing season of 1994 over eastern New Hampshire/southern Maine (WRS2 path 12/row 30). A near cloud free composite was created using five partly cloudy images acquired in that year, in which the scene acquired on 1994-07-01 serve as the base image. The acquisition year, month, and day of each image is represented by the 8 digits in the yyyy-mm-dd format.

FIGURE 9.5
Comparison of VCT disturbance maps produced using annual (top row) and biennial LTSS (bottom row) for an area in Alabama (WRS-2 path 21/row 37). The VCT was able to pick up many more partial canopy removals using an annual LTSS than using a biennial LTSS (middle row).

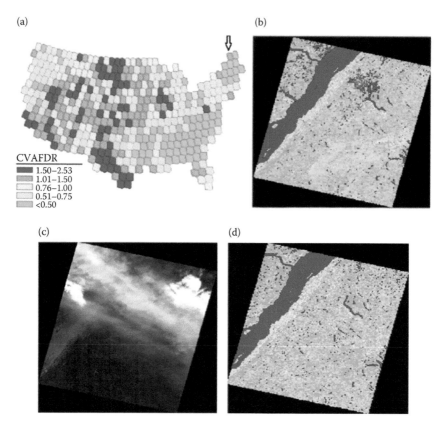

FIGURE 9.8

(a) 1999 nationwide cov(ADR) QA map. The arrow points to a potential analysis error (path 12/row 27) (b) The 1999 VCT annual disturbance map noting disturbance patterns not found in previous or later years (c). A true-color 1999 Landsat 5 composite image showing the presence of a smoke plume in this year. (d) VCT annual disturbance map after removal of misclassification errors.

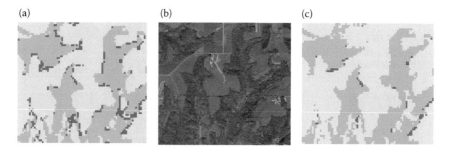

FIGURE 9.10

(a) A sample subset of annual forest disturbance map with misclassified disturbed pixels for path/row p25r29; (b) A subset of aerial photograph corresponding to (a); (c) the reclassified annual forest disturbance map.

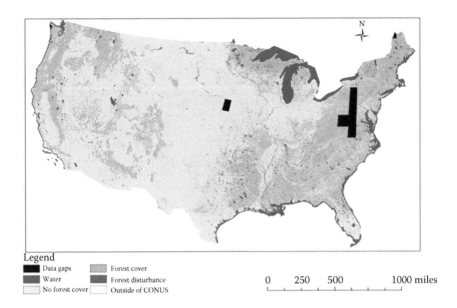

FIGURE 9.14
NAFD-NEX forest disturbance map for year 1998.

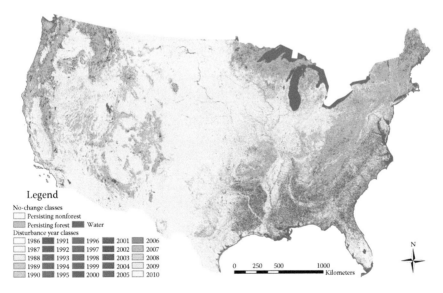

FIGURE 9.15
NAFD-NEX last integrated forest disturbance map.

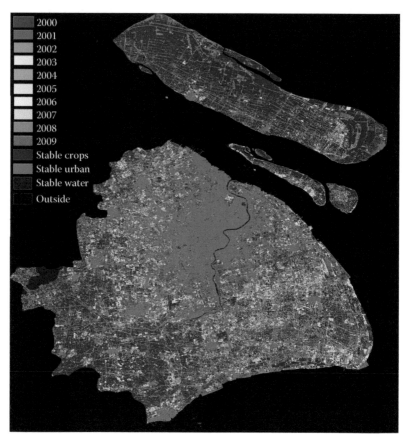

FIGURE 10.4
Annual agriculture land loss in Shanghai from 2000 to 2010 (Dark green: stable agriculture; Dark blue: stable water; Gray: stable urban area; Black: out of Shanghai).

FIGURE 10.5
Annual agriculture land loss in Shanghai from 2000 to 2010 in part of the Pudong district, Shanghai, China. Colors indicate times of conversion (Gray: stable land cover).

9

A General Workflow for Mapping Forest Disturbance History Using Pixel Based Time Series Analysis

Feng Zhao and Chengquan Huang

CONTENTS

9.1 Introduction 173
9.2 Overview of the NAFD-NEX Processing Flow 175
9.3 Image Selection and Preprocessing... 175
    9.3.1 Image Selection 176
    9.3.2 Image Preprocessing 177
    9.3.3 Image Compositing 177
9.4 VCT Disturbance Analysis 179
    9.4.1 Need for Annual Landsat Time Series 180
    9.4.2 Western US Sparse Forests Adjustment 182
9.5 Post Processing... 184
    9.5.1 Quality Assessment.. 185
    9.5.2 Adjustment of Minimum Mapping Unit (MMU)
            to Address Erroneous Forest Disturbance Rates
            in Low Forest Cover Counties.. 189
    9.5.3 Map Re-Projection 191
    9.5.4 Annual Disturbance Maps Mosaic... 192
9.6 NAFD-NEX Product Generation 192
    9.6.1 NAFD-NEX Product at Oak Ridge National Laboratory
            (ORNL)... 192
    9.6.2 Validation 195
9.7 Summary... 199
References.. 200

9.1 Introduction

Forests contain nearly 80% of the total carbon estimated to be the terrestrial aboveground biosphere, and thus it is critical to understand the role forests play in climate-change studies (Waring and Running et al., 1998). However, the magnitude of the forest carbon sinks and pools is uncertain, because disturbance and regrowth dynamics are not well characterized or understood.

Disturbance events, including harvest, fire, insect and storm damage, and disease, strongly affect carbon dynamics in many ways, including biomass removal, emissions from decaying biomass, and changes in productivity (Birdsey et al., 2013; King et al., 2007). Uncertainties related to disturbance and subsequent regrowth make it difficult to predict if forests will add carbon dioxide into the atmosphere or remove carbon dioxide from the atmosphere (King et al., 2007). To reduce these uncertainties, there is a need to develop temporal and spatial assessments, forest disturbance, and regrowth dynamics (Birdsey et al., 2009; Houghton & Goodale, 2004; Hurtt et al., 2002; Michalak et al., 2005; Tans, Fung, & Takahashi, 1990).

Remotely sensed data, specifically Landsat series of satellites, are being considered the best option for mapping forest-disturbance changes as Landsat observations offer longest historical archive (~40 years) of systematically collected remotely sensed data with a unique combination of a ~30 m spatial resolution and a temporal repeat frequency of 8–16 days (Goward et al., 2006; Hansen and Loveland, 2012; Cohen et al., 2017). Over the past several years, a variety of disturbance mapping algorithms have been developed, and there have been a palette of forest-disturbance products available for users of forest-disturbance products (Kennedy et al., 2010; Vogelmann et al., 2012; Huang et al., 2010b; Griffiths et al., 2014; Neigh et al., 2014; Hermosilla et al., 2016).

The main purpose of this chapter is to provide a processing workflow based on a forest-disturbance mapping project that can map forest disturbance at regional to global scales. It mainly focuses on the issues in Landsat data processing and data products that provide information on the occurrence, location, and timing of disturbance events. While it is important to understand the causal agents of disturbance events, which have become the focus of increasingly more studies (Hermosilla et al. 2015b; Kennedy et al. 2015; Moisen et al. 2016; Schroeder et al. 2011; Zhao et al. 2015), those studies are not discussed in this chapter.

The North American Forest Dynamics (NAFD) project, a core project of the North American Carbon Program (NACP), was designed to improve understanding of the North American carbon budget through use of Landsat time series and related US Forest Inventory Data (FIA) (Goward et al., 2008). The primary goal of NAFD was to reduce spatial and temporal uncertainty in the estimate of US forest dynamics over the last quarter century. NAFD began in 2003 with a prototype study and continued through two sampling phases. The results from these studies are reported (Masek et al., 2013) and the data products are available at the Oak Ridge National Laboratory Distributed Active Archive Center (ORNL DAAC) (Goward et al., 2015).

In phases I and II, Landsat data were sampled spatially (50 path/rows) and temporally (biennially) to reduce study costs, in particular, purchasing the Landsat data. The results from NAFD phases I and II found that disturbance estimates were nearly 20% lower than our reference validation results, indicating both spatial and temporal sampling had limited the mapping accuracy of the map products (Thomas et al., 2011).

The 2008 decision by the US Geological Survey to make the Landsat data archive available at no-cost to the user made a wall-to-wall, annual analysis of the conterminous United States (CONUS) feasible (Woodcock et al., 2008; Wulder et al., 2012). This comprehensive approach increased potential data volumes by at least a factor of 10. The processing volume issue was addressed through collaboration with the NASA Ames Research Center and their NASA Earth Exchange computing facility (*NEX*, https://nex.nasa.gov/nex/) (Nemani et al. 2011). As a result of this collaboration, the NAFD phase III wall-to-wall dataset is referred to as the NAFD-NEX data. These datasets are now also available at the ORNL DAAC (Goward et al., 2015). The new dimensions of data for the NAFD-NEX products required a new, automated approach that was a large step away from the analyst-dependent visual assessment, manual selection and processing approach used in NAFD phases I and II.

New challenges arose with wall-to-wall mapping at the national scale including automating image selection, compositing to overcome cloud-contaminated observations, achieving consistency across diverse landscapes, handling the large data volumes, and identifying and addressing quality assessment issues while under production, among others. This chapter focuses on these challenges, lessons learned and implications for future studies.

9.2 Overview of the NAFD-NEX Processing Flow

The NAFD-NEX disturbance mapping approach consisted of four major steps: (1) image selection and preprocessing, (2) Vegetation Change Tracker (VCT) disturbance analysis, (3) post processing, and (4) results validation (Figure 9.1). These steps reflect the requirements from acquiring images from USGS Earth Resources Observations and Science (EROS) to forming time series data, evaluating forest disturbance, production of twenty-five national 30 m disturbance maps, and validation of the resultant products.

9.3 Image Selection and Preprocessing

The quality of the scene selection and preprocessing are critically important to reduction of measurement errors in remote sensing change detection studies (Townshend et al., 2012). In NAFD Phase I and II, image selection was done visually and manually for each WRS2-path/row (Worldwide Reference System) locations (Masek et al., 2013), including selecting, processing, tracking, and assessing images through the radiometric and geometric processing

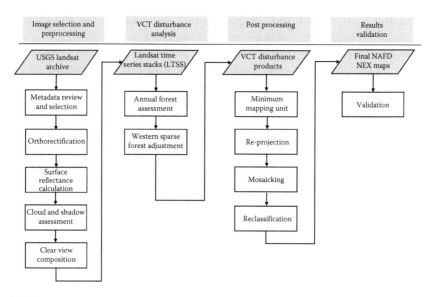

FIGURE 9.1
The workflow for producing NAFD-NEX nationwide annual forest disturbance maps.

steps, and proved extremely time intensive. In Phase III, we updated the
NAFD Image Selection and Processing (NISP) methods into an efficient,
automated, and repeatable stream with minimal analyst inputs (Figure 9.1).
Full details of the Phase III approach are available (Schleeweis et al., 2016). A
simple summary follows.

9.3.1 Image Selection

The goal for the NAFD-NEX processing stream was to compile one cloud free
(<5% cloud cover) forest disturbance map for the 434 scenes over the CONUS
for the time period 1984–2011. If cloud free images were available annually
during this 28 year time period then a total of ~12,152 Landsat scenes from
either Landsat 4, 5 or 7 would be used in the NAFD Image Selection and
Processing (NISP) approach.

All Landsat Thematic Mapper and Enhanced Thematic Mapper Plus (TM/
ETM+) US conterminous images available at the US Geological Survey (USGS)
Earth Resources Observation and Science (EROS) Center acquired between
1984 and 2011 for the 434 scenes were considered (~350,000 images). Filters
employed in NISP include Automatic Cloud Cover Assessment (ACCA)
cloud cover scores, peak growing season date ranges, Vegetation Change
Tracker (VCT) scene cloud contamination, scene size, and radiometry. In
67% (8193 scenes) of the cases, a single image per year was selected to meet
NAFD requirements. An additional 15,225 scenes were found suitable for
cloud compositing consideration. In 220 cases, no images were selected,
either because of gaps in the image archive or a paucity of quality images.

Generally, gaps were the result of Landsat operational problems (e.g., Landsat 4 TM technical problems or Earth Observation Satellite Company [EOSAT] acquisition strategies) or sensor malfunctions (i.e., failure of the scan line corrector mirror on Landsat 7 ETM+). These data gaps resulted in holes in the path/row annual time series (Schleeweis et al., 2016).

9.3.2 Image Preprocessing

Once all US Landsat images were selected, each image was preprocessed to provide the building blocks for the time series stacks. In NAFD Phase I and II, this preprocessing included (a) orthorectification, (b) radiometric calibration to Top-Of-Atmosphere (TOA) reflectance, (c) Surface Reflectance (SR) adjustment, (d) common scene area clipping, (e) cloud and shadow identification, and (f) cloud compositing (Thomas et al., 2011). For the Phase III study, the Landsat scene orthorectification was carried out at the EROS Center as a part of their Landsat level 1T processing stream (http://landsat.usgs.gov//Landsat_Processing_Details.php).

Radiometric calibration, TOA, and SR calculations were all conducted on NASA Earth Exchange (NEX) using the LEDAPS approach developed by Masek et al. (2006). Previous studies demonstrated that LEDAPS SR were highly consistent with MODIS SR (Feng et al., 2012, 2013). In this study, a random subset of results was visually inspected for errors (Schleeweis et al., 2016). Cloud contamination (clouds and associated shadows) in each selected scene was identified using the VCT cloud masking approach (Huang et al., 2010b). This approach creates a mask for each single image, flagging cloud, shadow, and other bad observations (e.g., missing scan lines, Landsat 7 SLC-off gaps, etc.) as well as clear view pixels. A normalized temperature band is also created for each input image by subtracting each pixel's Landsat brightness temperature from the mean temperature of forest pixels located within the same elevation zone as the target pixel. Forest pixels are identified using a dark object concept and Support Vector Machine approach (Huang et al., 2008).

9.3.3 Image Compositing

Clouds and their shadows as well as data gaps resulting from the Landsat 7 2003 scan line corrector failure create a need to composite images to obtain clear views. In this study, clear view Landsat scenes are defined as having less than 5% cloud cover.

We developed a clear view compositing (CVC) approach to take advantage of multiple images collected within the same growing season to produce clear view composites. In NAFD Phase III, for any WRS-2 path/row that did not have a single acceptable image (see Section 3.1) in a particular year, the CVC algorithm was applied to create a composite using two or more partly cloudy growing season images of that year. A flowchart of this algorithm is shown in Figure 9.2.

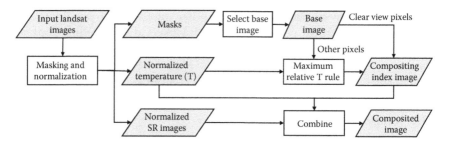

FIGURE 9.2
A flowchart of the clear view compositing (CVC) algorithm implemented in the VCT.

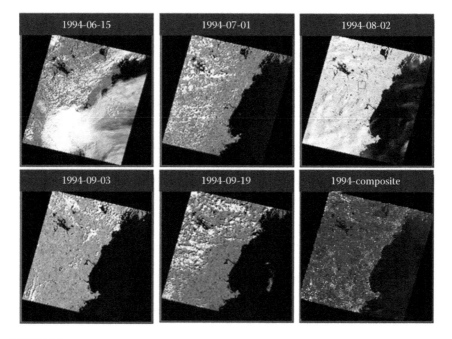

FIGURE 9.3
(**See color insert.**) No cloud free image was available during the summer growing season of 1994 over eastern New Hampshire/southern Maine (WRS2 path 12/row 30). A near cloud free composite was created using five partly cloudy images acquired in that year, in which the scene acquired on 1994-07-01 serve as the base image. The acquisition year, month, and day of each image is represented by the 8 digits in the yyyy-mm-dd format.

The CVC algorithm identifies a composite base image, which is the image with the most clear view pixels of the available images for that year and location (Figure 9.3). Reflectance bands of all other input images are normalized to the mean surface reflectance (SR) values of dark, dense forest (DDF) pixels in the base image (Huang et al., 2008). The remaining pixels are filled with the clear view pixels from the other available images using the observation that has the highest normalized temperature among all available. Because disturbed

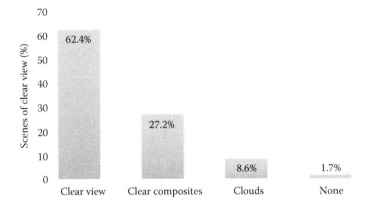

FIGURE 9.4
Following cloud compositing, the statistics of the images selected and produced for NAFD_NEX VCT process included 62% clear views, 27% composites, 9% cloud contaminated, and 2% no available scenes for the 424 oath/rows over the 28 year time period.

areas tend to have higher surface temperatures than undisturbed forests and other vegetated areas, this maximum normalized temperature compositing method is designed to preserve the disturbance signal over disturbed area, and hence may enhance disturbance detection by VCT.

An index image is created and used to select the values from the normalized input images. A quality flag mask is also created by combining the index image with the masks created by the cloud masking algorithm, which allows for tracking residual cloud/shadow and other bad observations that may still exist in the composited image, which can happen when no clear view observation exists. The CVC algorithm greatly increases clear view images in those years (e.g., Figure 9.3) without cloud free images.

Overall, the scene selection approach was only able to find 62% (7,581) of the scenes for NAFD NEX as clear view scenes (<5% cloud cover). The CVC process produced another 3301 (27%) composites with less than 5% residual cloud cover, for a total of ~90% of the needed scenes. Another 1048 (8.6%) of the needed scenes were identified or composited with cloud cover that varied from 5% to 50%. Finally, 202 (1.7%) of the needed scenes can either not be found or had major quality residual problems (e.g., excessive cloud) that made them not useable for NAFD NEX (Figure 9.4).

9.4 VCT Disturbance Analysis

The Vegetation Change Tracker (VCT) is an automated algorithm designed for evaluating forest disturbance and post disturbance recovery processes with spectral–temporal signals recorded in time series Landsat observations

TABLE 9.1

Annual Disturbance Map Class Definitions at WRS-2 Path/Row Level

Value	Name	Definitions
0	Background area	Area not covered by any valid land cover class
1	Persisting nonforest	Area not covered by trees in all the years
2	Persisting forest	Area covered by trees in all the years
4	Persisting Water	Water during the entire observing period
5	Forest recovered from disturbances	Area covered by trees, but was recovered from previous disturbances
6	Forest disturbance	Indicating the occurrence of a disturbance event over a forest area in the current year
7	Post disturbance nonforest	Area not covered by trees after disturbance

(Huang et al., 2009a,b; 2010). The VCT algorithm first identifies forest samples in each Landsat image, then based on these pixels, an integrated forest z-score (IFZ) index is calculated (Huang et al., 2010). Then the VCT algorithm tracks the changes of the IFZ series to determine whether and when forest disturbances occur. The VCT algorithm outputs scene-level annual forest disturbance maps, which consist of 8 classes (Table 9.1).

In NAFD phase III, we were able to address two main issues that had impacted the automation and accuracy of the NAFD phase I-II results: (1) clear view compositing (see Section 3.3) allowed for the use of high quality annual time series, which otherwise are plagued with too many poor quality pixels affected by clouds/shadow and SLC-off data gaps, (2) an objective way to handle the wide diversity of forest reflectance signals across ecological diverse regions, particularly in western US low density dry forest regions.

9.4.1 Need for Annual Landsat Time Series

NAFD phase I and II validation results revealed that minor disturbances, with partial canopy removal followed by rapid recovery, were not detected by the VCT with biennial image stacks (Thomas et al., 2011). We believed that many of those low magnitude disturbances, such as forest thinning, would be detected using annual Landsat Times Series Stacks (LTSS). To test this hypothesis, we built two versions of time series stacks, biennial and annual, and evaluated VCT results for each. In Washington state and North Carolina, disturbance products derived using annual LTSS in general had better accuracies than those derived using biennial LTSS (Huang et al., 2011, 2015).

To further evaluate these results, we visually inspected approximately 60 random pixel locations in Mississippi (p21/r37), North Carolina (p16/r35), Minnesota (p27/r27), and Washington (p47/r27). Locations were chosen for their diverse forest types and disturbance regimes. For each selected pixel, for "truth," the annual time series of Landsat imagery and high resolution

TABLE 9.2

Visual Assessment of Samples of Disturbances Mapped by Annual LTSS but Not by Biennial LTSS Randomly Selected from Four WRS2 Path/Rows

	21/37	16/35	27/27	47/27	Path/Row Total
Commission error by annual LTSS	8	1	10	5	23
Disturbance missed by biennial LTSS	36	37	37	19	129
Disturbance year in biennial results off by >5 years	15	21	12	33	82
Column Total	**59**	**59**	**59**	**57**	**234**

The value in each cell is the number of pixels evaluated.

images available from Google Earth were examined to determine whether disturbances had occurred.

This inspection revealed that more than 50% of the selected samples (129 out of 234) had true disturbances that were missed by the biennial VCT results (Table 9.2). Most of these plots appeared to experience minor canopy loss followed by rapid recovery, especially in the southern locations (p21/r37 and p16/r35) where forests may be thinned multiple times before the final harvest (Figure 9.5). Our conclusion is that more frequent repeat observations aid in the detection of these subtle events which, therefore, produce better accuracies.

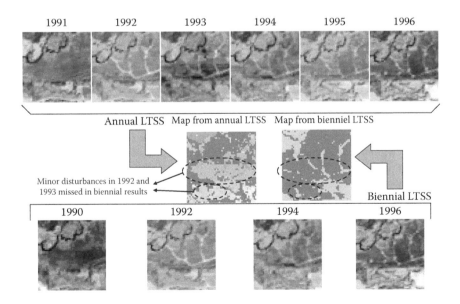

FIGURE 9.5
(**See color insert.**) Comparison of VCT disturbance maps produced using annual (top row) and biennial LTSS (bottom row) for an area in Alabama (WRS-2 path 21/row 37). The VCT was able to pick up many more partial canopy removals using an annual LTSS than using a biennual LTSS (middle row).

About one third of the plots (82 out of 234) had true disturbances detected in both the annual and biennial VCT results. In some of these plots, the disturbance year determined from biennial LTSS could be off by more than 5 years from the "truth" year of disturbance. Clouds in multiple sequential images over forest disturbance events appeared to be a main reason for the temporal difference. In both the phase I–II and phase III validation, a false positive is declared for a plot where the VCT year of detected disturbance is greater than ± 1 image year step from the "truth" year of disturbance. Therefore, an annual time series stack that is cloud cleared with the CVC algorithm can improve precision of the VCT year of disturbance and accuracy measures of the final map products. Only 10% (23 out of 234) of the selected samples had false positives in the annual VCT results.

9.4.2 Western US Sparse Forests Adjustment

The VCT leverages two indices through the time series to determine whether a pixel could possibly be "forestland." They are forest z-score (IFZ) index (Huang et al., 2009a) and normalized difference vegetation index (NDVI). In the eastern US where most of the forests have relatively dense canopy cover, IFZ and NDVI threshold values of 3 and 0.45 were found appropriate for separating forest from nonforested areas. In the western US, where forests have much higher IFZ values and lower NDVI values, this pair of eastern threshold values would result in forests with sparse density and low cover misclassified as nonforest.

During NAFD Phase I & II, IFZ-NDVI threshold values for low cover/ density conditions were determined iteratively through visual assessment of VCT results derived using different threshold values. To address the sparse forest issue, in NAFD Phase III, we developed a more systematic approach for calibrating the two threshold values to local forest conditions (at the Landsat scene level) using the USDA Forest Service Forest Inventory and Analysis (FIA) program field plot data. FIA plots are distributed across the US with a nominal interval of approximately 5 km (Smith, 2002). These plots are visited by field crews periodically to measure a suite of attributes. The FIA plot data used in this effort were collected with the annual plot design and comprised of both forested and nonforested FIA plots (Reams et al., 2005). Currently, FIA defines forested plots the same across all regions as those on lands that have or had (based on presence of stumps, snags, etc.) at least 10% crown cover and remain capable of producing a certain volume of timber, whereas nonforested plots are on lands with less than 10% canopy cover, including "other wooded land" with canopy cover up to 10%, or lands that have been removed from production (Smith et al., 2009).

We subset the FIA plots to create a reference dataset that mitigated definitional differences between FIA inventory data labels and the classification system used for VCT disturbance maps (hereafter called reference dataset). For the subset, plots with disturbance events recorded by either the VCT or FIA or both were excluded from the reference dataset. Plots with multiple FIA conditions of stand age, stand density, and multiple land covers (e.g., forest and nonforest) were also excluded. For example, FIA primarily recorded disturbances that occur in the same year as or several years prior to the field visit, whereas VCT may flag disturbance events that occur one year later (when a Landsat data gap prevents an image being available for that year or the image acquisition was prior to the event in the same year). Forest management activities such as thinning and harvest may be missed in the FIA disturbance category, but could be detected by VCT algorithm.

First, we assessed the suitability of using the IFZ-NDVI threshold values derived for close canopy eastern US forests on VCT analysis of dry sparse canopy western US forests by applying the IFZ and NDVI threshold values to WRS-2 path/rows located to the west of the 100° W longitude line. Overall agreement between the VCT classification and FIA reference dataset was calculated. As expected, the agreements were low for many of these western WRS-2 path/rows. The overall agreement appeared to be correlated with precipitation (Figure 9.6). WRS-2 path/rows with less than 75% agreement had an average annual precipitation of less than 400 mm, while those with greater than 93% agreement received on average more than 1200 mm precipitation annually (Figure 9.6). The lower agreements in the drier areas were generally

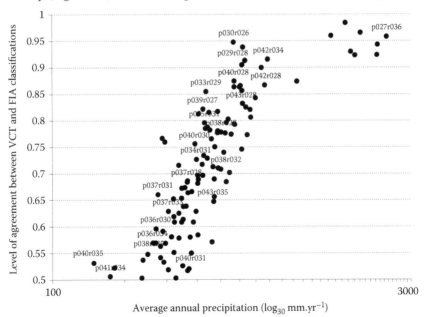

FIGURE 9.6
Level of agreement between FIA field records of single-condition forested/nonforested plots and VCT forest/nonforest classification produced using IFZ and VI thresholds of 3 and 0.45, respectively. The level of agreement (ranging from 0 to 1) was generally low in Landsat path/rows with low average annual precipitation.

TABLE 9.3

Error Matrix Showing Results from FIA Assessment for WRS-2 Path 37/Row 34

FIA/VCT	Forest	Nonforest	Total	User's Accuracy
Forest	59	386	445	13%
NonForest	3	422	425	99%
Total	62	808	870	
Producer's Accuracy	95%	55%		55%

VCT and FIA data included single-condition, nonforested, and undisturbed forested plots. The low level of agreement is mainly due to the high VCT omission errors of open canopy forests and woodlands.

due to forest class omission errors (the VCT incorrectly classifying forested pixels as nonforest pixels). Almost all omission errors were associated with plots that had very low biomass (<60 tons/ha). The number of VCT forested class commission errors were generally low (Table 9.3).

Next we used a Monte Carlo approach to determine the appropriate IFZ-NDVI threshold values for each of the western US (i.e., path/row located to the west of the 100°W longitude line) WRS-2 path/rows. The IFZ-NDVI threshold values varied from each other depending on the FIA reference dataset. Specifically, for each path/row, we randomly drew a pair of IFZ-NDVI threshold values, used them in the VCT forest/nonforest classification step, and then calculated the agreement between the new VCT classification and the FIA reference dataset. This was repeated 20,000 times for each WRS-2 path/row. The range of IFZ and NDVI thresholds was bounded between 0 and 30 for IFZ and 0 and 1 for NDVI. The pair of IFZ-NDVI threshold values that yielded the highest agreement between VCT and the FIA reference dataset for a WRS-2 path/row was selected as the optimal VCT threshold values for that path/row. Compared with the original threshold values optimized for eastern close canopy forests, these fine-tuned values resulted in increases of up to 40% in the agreement level between VCT and the FIA reference (Figure 9.7).

9.5 Post Processing

Once the VCT disturbance analysis was executed, several further evaluation and processing steps were need to complete development of the CONUS NAFD-NEX disturbance product. The first step is to conduct a quality assessment (QA) of the derived product. This QA process considered spatial patterns of the annual disturbance rates (Table 9.4) and the temporal variations in disturbance rates observed across the major US FIA sub regions. In the second step, we developed approaches to adjust the results to account for observed anomalies. In the third step, a minimum mapping unit was applied

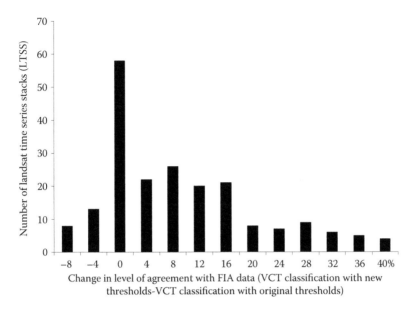

FIGURE 9.7
Frequency plot showing the number of Landsat Time Series Stacks with negative, no change, and positive changes in VCT classification level of agreement with FIA field data.

to the map results. The fourth post processing step merged the path/row results into a national map.

9.5.1 Quality Assessment

A number of stack level QA metrics were considered (Table 9.4). Based on an assumption that the amount of forest disturbances in neighboring stacks or in neighboring years should be closely related, if a metric (e.g., forest disturbance rate) for one year deviated significantly from its spatial or temporal neighbors, the stack was flagged for investigation of mapping commission or omission errors. Spatial neighbor anomalies were detected by visualizing the QA metrics per path/row in a national map (Figure 9.8). For example, one

TABLE 9.4

Statistical QA Metrics at Stack Level

Name of Metrics	Description
Annual Disturbance Rate ADR	Ratio of disturbed forest pixels to total forest pixels
Sum ($\sum$ADR)	Sum ADR—1986 and 2010
Mean (μADR)	Mean ADR—1986 and 2010
Standard Deviation (σADR)	Standard deviation ADR—1986 and 2010
Coefficient of Variation (cov(ADR))	Coefficient of variation ADR—1986 and 2010

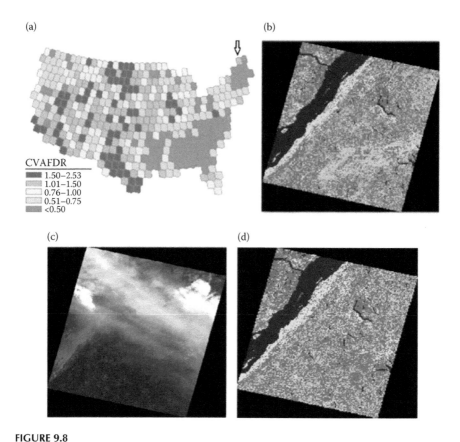

FIGURE 9.8
(**See color insert.**) (a). 1999 nationwide cov(ADR) QA map. The arrow points to a potential analysis error (path 12/row 27) (b) The 1999 VCT annual disturbance map noting disturbance patterns not found in previous or later years (c). A true-color 1999 Landsat 5 composite image showing the presence of a smoke plume in this year. (d) VCT annual disturbance map after removal of misclassification errors.

obvious potential outlier is in path/row 12/27. A significantly higher annual disturbance rate (ADR) is present in 1999 compared with its spatial neighbors (Figure 9.8a) suggesting that commission errors may have happened. Then we conducted a visual check and found that the larger disturbance rate in that year (Figure 9.8b) was actually a failure in the original VCT algorithm to identify clouds in that Landsat scene (Figure 9.8c). We subsequently improved the VCT cloud masking algorithm, and applied the updated cloud masking algorithm to all VCT stacks such that it successfully masked all clouds and removed the false disturbance in our final product (Figure 9.8d). This QA approach helped focus analyst attention on map products for which costly visual assessments would be most beneficial.

Besides the cloud issue, we have also identified an issue of regional anomalies in the North Central US based on the standard deviation of forest disturbance rate. Change detection in image time series can be complex as the borders between classes are shifting through time. Using our QA approach we identified two regions that suffered from this problem; the North Central US and the State of Florida (Figure 9.9). In both cases shifts occur as a result of variable moisture conditions, primarily droughts.

Specifically, in the North Central US, we found large disturbance rates in 1987–1988 covering 129 path/rows (Figure 9.9). These can be attributed to a widespread drought in the region (Andreadis et al., 2005). This drought caused a shift in the spectral boundary between forests, agriculture, and grasslands (Figure 9.10a). Figure 9.10a shows a subset of a forest disturbance map located in southern Minnesota where landscape is dominated by agricultural crops (Figure 9.10b). Compared with forest stand and agricultural crops, herbaceous grasses are relatively vulnerable to drought because they are considered sensitive to microclimate (De Frenne et al., 2011). Similarly we found anomalous disturbance patterns throughout the State of Florida, primarily in the wetlands areas. In these locations the errors were caused by mixed forest, wetland, and water cover along rivers and streams which varied from year to year as water levels changed.

To handle these false positive disturbance measurements, National Land Cover 1992 dataset (NLCD 1992) (Vogelmann et al., 2001) was used to remove

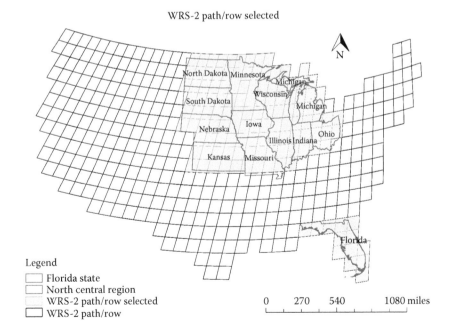

FIGURE 9.9
The 434 WRS-2 path/row needed to provide a contiguous coverage of CONUS. The shaded path/rows in north central region and Florida State were post processed primarily for removal of false positive forest disturbance results in agricultural and wetlands landscapes due to drought.

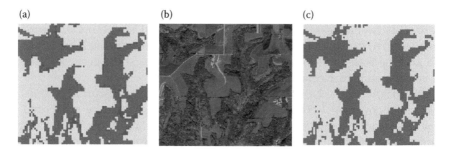

FIGURE 9.10
(See color insert.) (a). A sample subset of annual forest disturbance map with misclassified disturbed pixels for path/row p25r29; (b) A subset of aerial photograph corresponding to (a); (c) the reclassified annual forest disturbance map.

those misclassified disturbances. Specifically in the North Central region, any disturbance pixel observed in VCT in years 1987 and 1988 that occurred within a 3 × 3 neighborhood of planted/cultivated class categories in NLCD 1992 was converted to nonforest class. Meanwhile, disturbance pixels that were classified as emergent herbaceous wetland in NLCD 1992 were converted to nonforest class. A similar approach was employed in Florida where we changed the disturbance pixels which were defined as emergent herbaceous wetland in VCT into nonforest class. Although the potential misclassification in NLCD 1992 can introduce removal of true disturbances or may not remove some disturbances, the application of NLCD 1992 would remove false disturbances only, and the errors are limited in North Central CONUS and Florida only.

Figure 9.11 shows an example that demonstrated the forest disturbance rate change after removing those false disturbance measurements. The sample

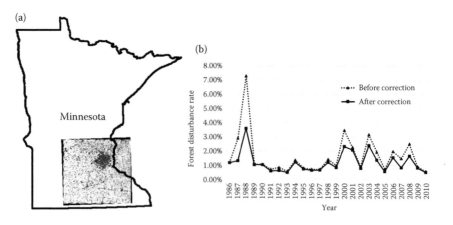

FIGURE 9.11
(a) Location of p27/r29; (b) a comparison of forest disturbance rate for p27/r29 before correction and after correction.

path 27/row 29 location was selected as it was one stack that has experienced the drought in 1987 and 1988, and also the area it covers is dominated by a mosaic of Planted/Cultivated class (66.9%), Forest (11.5%), and Emergent herbaceous wetland (3.2%) based on the NLCD 1992. The change of forest disturbance rates in year 1987/1988 suggested the correction did successfully remove a large amount of false disturbance detection along the border of forest and Planted/Cultivated area. Meanwhile, the false forest disturbance measurements in the emergent herbaceous wetland were corrected.

9.5.2 Adjustment of Minimum Mapping Unit (MMU) to Address Erroneous Forest Disturbance Rates in Low Forest Cover Counties

In Phase I&II, a spatial-temporal MMU filter was applied to the time series of annual maps; however, the values in the annual map products were not harmonized through the time series after the MMU was applied (Thomas et al., 2011). Harmonization between the time steps of the filtered map series is needed to ensure continuity in classes that by definition imply continuity (i.e., persistent classes [persistent forest, persistent nonforest, and persistent water] are consistent throughout the entire time series; Disturbance classes, [including forest recovered from disturbance, forest disturbance, and post disturbance nonforest] have no persistent classes throughout the entire time series). This is particularly relevant for time series maps as temporal consistency is essential to understand the subtle changes in a short time interval.

While examining the national annual disturbance rates, we observed improbably high rates of disturbance, typically in the Great Plains region. On examination we found these rates were driven by the accumulation of many small errors related to small forest patches and their spurious false positive disturbances, due to misregistraion and pixel mixing. To address where these errors were accumulating, we first identified counties that had less than 6% forest cover and less than 150 km² total forest area based on National Land Cover 1992 dataset (NLCD 1992) (Vogelmann et al., 2001). We then stratified the annual mosaicked NAFD maps into two strata, forested counties and low forest cover (LFC) counties (Table 9.5), and applied a larger MMU threshold in the LFC counties strata so that potential commission errors due to misregistration and pixel mixing can be removed (Figure 9.12).

TABLE 9.5

MMU Size for NAFD-NEX Product

VCT Class	MMU for Forested Counties Region (Number of Pixels)	MMU for Low Forest Cover (<6%) Counties Region (Number of Pixels)
Persistent forest	2	2
Persistent nonforest	2	2
Disturbed patch	4	10

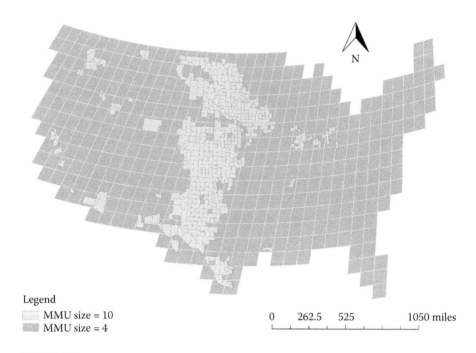

Legend
MMU size = 10
MMU size = 4

0   262.5   525   1050 miles

FIGURE 9.12
WRS-2 path/row with different forest disturbance patch MMU size. Static classes were treated the same in all WRS-2 path/row, with a MMU of 0.18 ha.

The MMU approach uses spatial and temporal neighbor rules to define minimum thresholds which vary by strata. The MMU size for low forest cover region is different from that of other regions (Figure 9.13) for adjustment. To calculate patch sizes, a 9-neighbor spatial rule was used for both strata and all map classes. However, for disturbed forest patches, we used different temporal neighbor rules in the two strata.

In the forested counties strata, we used a 3 year neighbor rule, so all disturbed forest pixels in a moving three year window (map year ±1 map year) were included in the temporal "patch." In LFC counties, a more conservative one year neighbor rule was used, so only pixels disturbed in that year were included in the patch. Also, in LFC counties, a larger spatial MMU threshold of 9 pixels (0.81 ha) was applied than in forested counties which had a four pixels (0.36 ha) spatial MMU threshold for disturbed forest patches. In both strata, a spatial MMU of two pixels (0.18 ha) was used for static classes, such as water, undisturbed forest, and nonforest. Finally, we applied a temporal adjustment to ensure the consistency of annual disturbance maps in the temporal domain. The goal of temporal adjustment was to avoid the errors introduced by MMU analysis.

Pixels below the MMU thresholds had their values adjusted by decision rules that favored static classes. No new forest disturbance was introduced

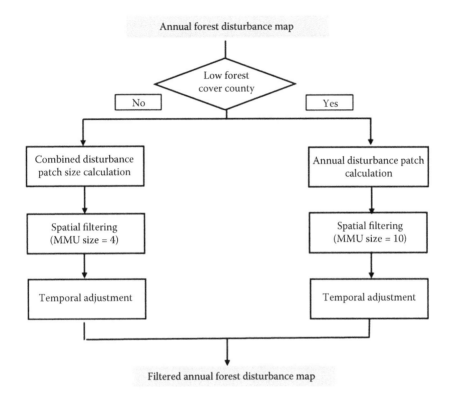

FIGURE 9.13
A flowchart of MMU filtering algorithm for the class of disturbed forest pixels as adapted to address erroneous forest disturbance rate evaluation on low forest cover in US counties.

by the MMU filtering, as we reclassified pixels only to static class values. For example, for a pixel with only one disturbance record in its original time series of annual values, and if it fell below the MMU thresholds, its value (in all of the annual maps in the time series) would be changed to a static class. If the rest of the time series value for that pixel are forests in the year with the disturbed value it is replaced by persisting forest class.

9.5.3 Map Re-Projection

The USGS level 1T products are produced in the Universal Transverse Mercator (UTM) projection (Snyder, 1987). This is not suited to the product of an integrated national map as the UTM projection for the US consists of 10 different UTM zones. We employed the Albers Equal Area projection for the integrated national map and annual disturbance maps, a widely adopted projection for several Landsat-based products (e.g., Web-enabled Landsat Data (WELD)) (Roy et al., 2010) and National Land Cover Database (NLCD) (Vogelmann et al., 2001).

The UTM files were re-projected to Albers projection using the inverse gridding approach and nearest neighbor resampling approach based on the General Cartographic Transformation Package (GCTP) developed by the USGS.

9.5.4 Annual Disturbance Maps Mosaic

Because the original Landsat path/rows overlap, it is necessary to mosaic the observations for adjacent path/rows to produce a consistent national map. The NAFD CONUS grid has the exact dimensions of the NLCD 1992–2001 retrofit data product (97,646 rows and 154,320 columns). Each pixel in the annual UTM path/row maps was directly assigned into the national grid. The mosaicking process started assigning pixels to the upper left corner of the national grid along the path direction. In the overlapping regions, pixels in the lower latitude path/rows were retained over those in higher latitude path/rows. When all the pixels in a single path were assigned, we then mosaicked the next path to the right, and consequently, pixels in the next path would be retained over the previous path.

Although the NAFD-NEX study examined Landsat data from 1984 to 2011, only disturbance results from 1986 to 2010 are provided. The 1984 and 2011 maps are not included due to the nature of the VCT time series algorithm which suffers in the first and last year of the time series with pixels flagged as bad observations which cannot be filled by the CVC compositing algorithm (see Section 3.3).

There were also substantial data gaps in both 1984 and 1985. Data gaps were found in later years as well (Figure 9.14) as a result of either missing data or persistent cloud contamination. In addition, approximately 0.13% of the CONUS land area was not included in the NAFD mosaics. Two additional path/rows would have been needed, but were not included in the NAFD-NEX study due to their small amount of land area.

9.6 NAFD-NEX Product Generation

9.6.1 NAFD-NEX Product at Oak Ridge National Laboratory (ORNL)

The NAFD-NEX disturbance products have been made available through the ORNL North American Carbon Project (NACP) Distributed Active Archive Center (Goward et al., 2015) (http://dx.doi.org/10.3334/ORNLDAAC/1290). These NAFD-NEX products include two sets of national mosaics. The first set includes 25 annual maps, with each flagging disturbances observed in a year between 1986 and 2010 as well as several other classes (Table 9.6). These

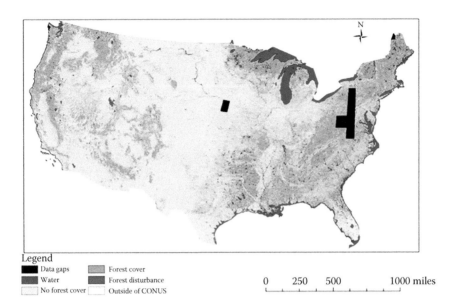

Legend

■	Data gaps	▨	Forest cover
■	Water	▨	Forest disturbance
▢	No forest cover	▢	Outside of CONUS

0   250   500   1000 miles

FIGURE 9.14
(See color insert.) NAFD-NEX forest disturbance map for year 1998.

maps can be used to calculate annual disturbance rates as well as forest cover in each year.

The second set of products provides a comprehensive forest disturbance history by integrating the annual maps (Figure 9.15). These maps provide 1986–2010 synoptic views of the spatial–temporal dynamics of forest disturbance across the CONUS for the first and last disturbances (Table 9.7). Since substantial portions of the forests in the US experienced only one disturbance between 1986 and 2010, these two maps are identical in areas where one or no disturbances were detected during the entire observing period.

TABLE 9.6

Annual Disturbance Map Class Definitions for National Mosaic

Value	Name	Definitions
0	Data gaps	No acceptable data available.
1	Water	Area covered by water during the entire observing period
2	No forest cover	Area not covered by trees in the current year but may or may not have tree cover in other years during the observing period
3	Forest cover	Area covered by trees in the current year but may or may not have tree cover in other years during the observing period
4	Forest disturbance	Indicating the occurrence of a disturbance event over a forest area in the current year
254	Outside of study area	Area outside the study area

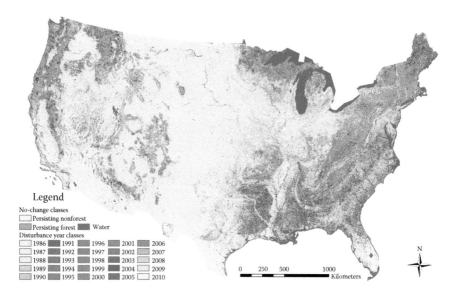

FIGURE 9.15

(**See color insert.**) NAFD-NEX last integrated forest disturbance map.

TABLE 9.7

Class Definitions for Time Integrated Disturbance Map Mosaic

Value	Name	Definitions
Stable Classes		
1	Water	Area covered by water during the entire observing period
2	Nonforest	Area never covered by trees during the entire observing period
3	"Undisturbed" forest	Forested area that had no disturbance detected during the entire observing period. This includes areas that (1) had tree cover throughout the entire observing period, (2) were disturbed before the beginning of the time series, but returned to forest cover and were not disturbed again, and (3) were nonforest at the beginning of the observing period but had tree cover in later years (e.g., woody encroachment, agricultural abandonment, urban tree growth).
254	Outside of study area	Area outside the study area
Disturbance Classes		
15–40	Disturbance year	Area disturbed in a particular year. Disturbance year = 1970 + class value. Only show the latest disturbance if an area was disturbed more than once during the observing period. The mapped disturbance event typically occurred between the acquisition dates of two Landsat images—one acquired in the disturbance year and the other in the immediately previous year, that were used in the VCT analysis.

9.6.2 Validation

The spatial and temporal dimensions of the NAFD-NEX products add new challenges in both the sampling and response design for validation of forest change map products. While validation of phase I & II results used an opportunistic approach to quantify mapping errors in different forest biomes and disturbance regimes (Thomas et al., 2011; Masek et al., 2013), a probability based sampling method was used to select samples for validating the NAFD-NEX products and for determining potential disturbance area estimation biases arising from mapping errors (Olofsson et al., 2014). Specifically, a two stage stratified random cluster design was used to draw a higher proportion of plots from forested regions. During the first stage, the boundaries among the WRS-2 path/rows needed to cover CONUS were redrawn to create Thiessen Scene Area (TSA) polygons such that there were neither gaps nor overlaps between adjacent TSA polygons. These TSAs were divided into five strata based on a combination of ecoregions and forest type distribution (Omerik, 1987; Ruefenacht et al., 2008). In each stratum, the number of TSAs to be selected was proportional to the product of the area of that stratum and the total forest area in that stratum. A total of 180 out of 434 TSAs were selected. During the second stage, a simple random sampling method was used to select 40 plots, or 30-m Landsat pixels, within each selected TSA. This resulted in a total of 7,200 plots over CONUS, a sample size that was evaluated to available resources and was still large enough to yield sufficiently small sampling errors.

For each selected plot, reference data was collected using TimeSync, a software package designed for visual interpretation of land cover and land use change at annual time steps based on Landsat time series, high resolution imagery available in Google Earth and ancillary datasets (Cohen et al., 2010). The land cover class for each plot was recorded through time. Years with visual evidence of change were noted with a corresponding change process. Spectral indices were used along with training data to model annual canopy cover. Full details on the reference data including details on the Landsat time series stacks used has been provided by Cohen et al. (Cohen et al., 2016). It is important to note that the image acquisition dates and cloud compositing approaches for the NAFD-NEX time series and the reference data collection were independent.

A number of factors were considered in determining a disturbance year match between NAFD-NEX and the reference data. For abrupt changes that had short durations, the disturbance years from the two datasets were deemed to match each other if their differences were 1 year or less. Here the 1-year buffer was factored in to account for difficulties in determining the "true" disturbance year using Landsat data, which can arise from use of images with different acquisition dates, obscured views due to cloud, shadow, SLC-off gaps, and so on, subjective analyst interpretations, and differing magnitude definitions. For gradual changes that occurred over multiple years (e.g., forest decline), disturbance year was recorded differently between NAFD-NEX and TimeSync. For a gradual disturbance that occurred over N years, TimeSync typically recorded N events, one in each of the N years. But in NAFD-NEX, this disturbance was considered a single event recorded in the year when a predefined threshold value was reached (Huang et al., 2010). To minimize the impact of such definition differences on the validation results, a NAFD-NEX disturbance year value was considered to agree with the reference data if it was within the disturbance year range of a gradual disturbance identified in the reference data.

While VCT was designed to detect disturbances that caused substantial canopy cover loss, in deriving the reference data, efforts were made to flag all disturbances that resulted in changes identifiable by the image analysts using high resolution aerial imagery and Landsat in TimeSync (Cohen et al., 2016). As a result, the Cohen et al. (2016) reference dataset is sensitive to a wide variety of forest canopy disturbances, wider than might be expected using Landsat imagery alone, regardless of the algorithm. To assess the sensitivity of the NAFD-NEX product to disturbance intensity, we gradually increased the disturbance intensity threshold value from 0% to 100% in flagging disturbances in the reference data and recalculated the commission and omission errors at each threshold value. An approximate balance between the two errors was reached at a disturbance intensity threshold value of ~20%, suggesting that most of the disturbances identified in the NAFD-NEX product had a minimum canopy cover loss of 20% (Figure 9.16). The accuracies reported for the NAFD-NEX product were derived by using this threshold value to determine forest disturbances in the reference samples. Disturbances identified by TimeSync with disturbance intensities less than 20% were recoded to undisturbed forest. Of the disturbance records affected

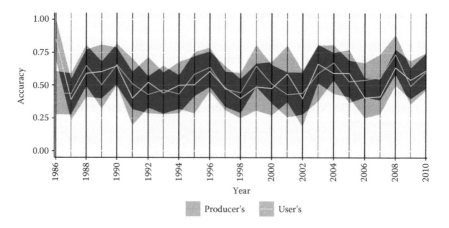

FIGURE 9.16
Producer's and user's accuracy for disturbance class with 95% confidence envelope shown in gray. Error assessment was repeated using a different threshold for disturbance magnitude in the reference data using more than 20% absolute canopy loss (per forest disturbance event).

by this reassignment, 60% were classified as harvest in the reference dataset, and only 25% were in the stress category. The reference dataset did not have enough plots to reliably (reasonably low standard errors) report these findings by region.

Accuracy measures, including overall accuracy, and class specific user's and producer's accuracies, as well as biases arising from mapping error, were calculated following standard methods (McRoberts and Walters, 2012; Stehman, 2013; Olofsson et al., 2014). When disturbance year match was not considered, the NAFD-NEX product had an overall accuracy of 84.6% with three classes: disturbed forest, undisturbed forest, and nonforest (Figure 9.16). The user's and producer's accuracies of the disturbed forest class were 62.9% and 66.9%, respectively (Tables 9.8 and 9.9). The individual year maps do not distinguish persisting forest or nonforest (see Table 9.5). At the individual year level, water, no forest cover, and forest cover were grouped into a single "other" class in accuracy calculation, because definitional differences regarding forest regrowth prevented reliable separation. Ultimately, regrowth is a slow process that is difficult to assign to a single year, and the Cohen et al. (2016) dataset was not designed to validate a "year of regrowth." Having multiple analysts visually assess in what year a pixel had returned to 10% canopy cover in a TM image without the aid of high resolution imagery, which is not available for the earlier part of the Landsat record, would be unreliable. VCT regrowth has been investigated in Zhao et al. (2016) and includes a deeper discussion of measuring and validating regrowth. The overall accuracies of the individual year maps were all above 90%. The disturbance class had user's accuracies ranging from 42.8% to 73.6% and producer's accuracies from 40.0% to 84.8% (Figure 9.16). However, the area of disturbed forest mapped by VCT in each year was not statistically different from the area of disturbed forest estimated by the reference data, meaning there was no statistically significant mapping bias in all years, except 2007. In the 2007 annual map of forest disturbance, the VCT class of disturbed forest was 520 km^2 larger than disturbed forest reference data (using 95% confidence bounds around the estimate).

TABLE 9.8

Error Matrix for Overall Map Classes, Represented as Proportion of Total Land Area Mapped

	Reference					
VCT Mapped	**Non Forest**	**Persisting Forest**	**Disturbed Forest**	**SUM**	**User Accuracy**	**Standard Error ($\pm$)**
Non Forest	0.604	0.069	0.07	0.68	0.888	0.015
Persisting Forest	0.023	0.186	0.021	0.23	0.809	0.015
Disturbed Forest	0.008	0.026	0.057	0.09	0.629	0.014
SUM	0.634	0.281	0.085	0.84648		
Producer Accuracy	0.952	0.662	0.669			
Standard error ($\pm$)	0.007	0.025	0.021			

TABLE 9.9

Error Matrix in Proportion of Area

	NF	PF	1986	1987	1988	1989	1990	1991	1992	1993	1994	1995	1996	1997	1998	1999	2000	2001	2002	2003	2004	2005	2006	2007	2008	2009	2010	Commission	Standard Error (±)
NF	**58.23**	6.81	0.03	0.05	0.02	0.05	0.03	0.06	0.04	0.06	0.02	0.02	0.00	0.00	0.06	0.04	0.02	0.01	0.02	0.04	0.01	0.04	0.04	0.01	0.02	0.02	0.04	0.12	0.02
PF	2.18	**18.45**	0.08	0.11	0.09	0.08	0.05	0.04	0.13	0.06	0.08	0.08	0.09	0.11	0.07	0.04	0.12	0.18	0.10	0.06	0.04	0.06	0.18	0.09	0.08	0.15	0.06	0.19	0.01
1986	0.00	0.06	**0.22**																									0.21	0.06
1987	0.06	0.22		**0.22**																								0.56	0.08
1988	0.03	0.12		0.01	**0.23**																							0.46	0.10
1989	0.00	0.15				**0.23**																						0.41	0.10
1990	0.02	0.14				0.01	**0.35**																					0.36	0.07
1991	0.04	0.19						**0.18**																				0.57	0.12
1992	0.04	0.13							**0.15**	0.02																		0.60	0.09
1993	0.04	0.12								**0.20**																		0.45	0.09
1994	0.01	0.10									**0.19**	0.00																0.42	0.09
1995	0.03	0.20										**0.18**																0.56	0.13
1996	0.02	0.15											**0.22**															0.50	0.08
1997	0.03	0.16												**0.21**														0.48	0.10
1998	0.01	0.24													**0.20**													0.58	0.08
1999	0.01	0.24														**0.28**												0.53	0.13
2000	0.01	0.11														0.00	**0.23**	0.01										0.55	0.12
2001	0.06	0.18																**0.27**										0.39	0.09
2002	0.12	0.19																	**0.15**									0.67	0.11
2003	0.03	0.17																		**0.22**								0.53	0.09
2004	0.05	0.06																			**0.24**							0.49	0.08
2005	0.01	0.44																				**0.22**						0.29	0.08
2006	0.01	0.19																					**0.27**					0.67	0.10
2007	0.06	0.20																					0.02	**0.27**				0.53	0.08
2008	0.02	0.10																							**0.31**			0.43	0.08
2009	0.05	0.10																								**0.31**		0.35	0.07
2010	0.02	0.12																									**0.17**	0.47	0.10
Omission	0.05	0.37	0.37	0.42	0.33	0.39	0.25	0.37	0.54	0.41	0.43	0.39	0.36	0.44	0.45	0.28	0.42	0.44	0.58	0.44	0.27	0.36	0.45	0.30	0.26	0.39	**0.82**		
Standard Error (±)	0.01	0.02	0.09	0.08	0.08	0.09	0.08	0.10	0.07	0.10	0.08	0.09	0.08	0.07	0.08	0.07	0.09	0.08	0.12	0.07	0.08	0.07	0.09	0.08	0.08	0.08	0.08		

NF = nonforest, PF = persisting forest, or forest not disturbed during the time series. 1986–2010 reflect the years of the time series and numbers in these columns reflect the year of onset of disturbance in the reference dataset, again by proportion of land area mapped. SE is one standard error from the probabilistic sampling design of the reference dataset.

9.7 Summary

A comprehensive forest disturbance history is essential to understand the impact of forest disturbance history on regional carbon budget (Thornton et al., 2002; Williams et al., 2014). With a detailed characterization of both the spatial distribution of forest disturbances over CONUS, and the 25 years' history of forest disturbance, the NAFD-NEX product provides a unique opportunity for regional carbon studies.

This study suggests that annual image time series hold more information on subtle disturbances, such as thinning, partial harvest, and mortality from insects and disease, than biennial time series allowing for more accurate detection of these important processes. Since post disturbance recovery varied with intensity and type of treatment, the impact of these low magnitude disturbances may increase future carbon uptake, resilience to high severity fire, and climate stresses (Dore et al., 2016). As a result, future studies on mapping the forest disturbance intensity and disturbance type and evaluating the impact of low magnitude disturbances on regional carbon budget would be critical to regional carbon studies.

Pixel-based approaches, such as best pixel approach (Griffiths et al., 2014; White et al., 2014; Hermosilla et al., 2016), rather than scene-based approaches help alleviate gaps in the Landsat archive due to cloud contamination whether the approach uses all images (Zhu et al., 2015) or only all images throughout the growing season (Hansen et al., 2013; White et al., 2014). Currently, as the VCT approach is being adapted to handle Landsat 8, its compositing approach is also being modified to take advantage of scene overlap areas to reduce the impact of cloud/shadows and bad/missing data in the Landsat record (Holden & Woodcock, 2016).

Incorporating multi-season imagery into the masking phase of VCT has shown advantages in certain locations for removing spurious changes at wetland-agriculture-forest intersections. Future Landsat missions (e.g., Landsat 9, 10, etc.) will need to consider further advances in temporal coverage along with merging observations from other observation systems, such as the European Sentinel-2 system.

All current VCT processing steps were applied only to Landsat 5 TM and Landsat 7 ETM+ data. The VCT algorithm is currently being updated to take advantage of the Landsat 8 Operational Land Imager (OLI) to update the forest disturbance history to 2016. Considerable further work would be needed to incorporate the earlier 1972–1992 Landsat 1–5 Multispectral Scanner System (MSS) observations because of its spatio-temporal characteristics that differ substantially from that of the Landsat 4–8 satellites.

Sparse forest adjustments based on FIA reference dataset is critical for accurate mapping of forest disturbances in the western CONUS. The threshold varied for each LTSS, and can be used repeatedly for forest disturbance mapping in the western CONUS when additional images are added for updating forest disturbance history maps.

This study also provide a practical solution to assess the product with a series of summary metrics to adjust the parameter to handle sparse forest in the US west, and to remove anomalies in the low forest cover region. The solution improved the classification accuracies by taking into account local forest conditions.

The general workflow used in producing NAFD-NEX annual forest disturbance history product provides a pathway to consider automated large area processing of Landsat time series data for many land cover change analyses beyond forest disturbance analysis. Using the NEX computing environment, the North American Forest Dynamics (NAFD) Project was able to process 80,000+ Landsat images (L5-L7), creating estimates and annual maps of forest history to reduce uncertainty in the spatial and temporal dynamics of forest change for the CONUS (Nemani et al., 2011).

References

Andreadis, K.M., Clark, E.A., Wood, A.W., Hamlet, A.F., Lettenmaier, D.P., 2005. Twentieth-century drought in the conterminous United States. *J. Hydrometeorol.* 6, 985–1001. doi:10.1175/JHM450.1

Birdsey, R., Bates, N., Behrenfeld, M., Davis, K., Doney, S.C., Feely, R., Hansell, D., Heath, L., Kasischke, E., Kheshgi, H., Law, B., Lee, C., McGuire, A.D., Raymond, P., Tucker, C.J., 2009. Carbon cycle observations: Gaps threaten climate mitigation policies. *Eos (Washington. DC).* 90, 292. doi:10.1029/2009EO340005

Birdsey, R., Pan, Y., Houghton, R., 2013. Sustainable landscapes in a world of change: Tropical forests, land use and implementation of REDD+: Part II. *Carbon Manag.* 4, 567–569. doi:10.4155/cmt.13.67

Cohen, W.B., Healey, S.P., Yang, Z., Stehman, S. V., Brewer, C.K., Brooks, E.B., Gorelick, N. et al., 2017. How similar are forest disturbance maps derived from different Landsat time series algorithms? *Forests* 8. doi:10.3390/f8040098

Cohen, W.B., Yang, Z., Kennedy, R., 2010. Detecting trends in forest disturbance and recovery using yearly Landsat time series: 2. TimeSync—Tools for calibration and validation. *Remote Sens. Environ.* 114, 2911–2924. doi:10.1016/j.rse.2010.07.010

Cohen, W.B., Yang, Z., Stehman, S. V., Schroeder, T.A., Bell, D.M., Masek, J.G., Huang, C., Meigs, G.W., 2016. Forest disturbance across the conterminous United States from 1985–2012: The emerging dominance of forest decline. *For. Ecol. Manage.* 360, 242–252. doi:10.1016/j.foreco.2015.10.042

De Frenne, P., Brunet, J., Shevtsova, A., Kolb, A., Graae, B.J., Chabrerie, O., Cousins, S.A. et al., 2011. Temperature effects on forest herbs assessed by warming and transplant experiments along a latitudinal gradient. *Glob. Chang. Biol.* 17, 3240–3253. doi:10.1111/j.1365-2486.2011.02449.x

Dore, S., Fry, D.L., Collins, B.M., Vargas, R., York, R.A., Stephens, S.L., 2016. Management impacts on carbon dynamics in a sierra Nevada mixed conifer forest. *PLoS One* 11(2). doi:10.1371/journal.pone.0150256

Feng, M., Huang, C., Channan, S., Vermote, E.F., Masek, J.G., Townshend, J.R., 2012. Quality assessment of Landsat surface reflectance products using MODIS data. *Comput. Geosci.* 38, 9–22. doi:10.1016/j.cageo.2011.04.011

Feng, M., Sexton, J.O., Huang, C., Masek, J.G., Vermote, E.F., Gao, F., Narasimhan, R., Channan, S., Wolfe, R.E., Townshend, J.R., 2013. Global surface reflectance products from Landsat: Assessment using coincident MODIS observations. *Remote Sens. Environ.* 134, 276–293. doi:10.1016/j.rse.2013.02.031

Goward, S., Arvidson, T., Williams, D., Faundeen, J., Irons, J., Franks, S., 2006. Historical record of Landsat global coverage. *Photogramm. Eng. Remote Sens.* 72, 1155–1169. doi:10.14358/PERS.72.10.1155

Goward, S.N., Huang, C., Zhao, F., Schleeweis, K., Rishmawi, K., Lindsey, M., Dungan, J.L., Michaelis, A., 2015. *NACP NAFD Project: Forest Disturbance History from Landsat, 1986–2010.* In University of Maryland NAFD Team (Ed.). College Park MD: ORNL DAAC, Oak Ridge, Tennessee, USA.

Goward, S.N., Masek, J.G., Cohen, W., Moisen, G., Collatz, G.J., Healey, S., Houghton, R.A. et al., 2008. Forest disturbance and North American carbon flux. *Eos, Trans. Am. Geophys. Union.* 89, 105–106.

Griffiths, P., Kuemmerle, T., Baumann, M., Radeloff, V.C., Abrudan, I. V., Lieskovsky, J., Munteanu, C., Ostapowicz, K., Hostert, P., 2014. Forest disturbances, forest recovery, and changes in forest types across the carpathian ecoregion from 1985 to 2010 based on landsat image composites. *Remote Sens. Environ.* 151, 72–88. doi:10.1016/j.rse.2013.04.022

Hansen, M.C., Loveland, T.R., 2012. A review of large area monitoring of land cover change using Landsat data. *Remote Sens. Environ.* 122, 66–74. doi:10.1016/j.rse.2011.08.024

Hansen, M.C., Potapov, P.V., Moore, R., Hancher, M., Turubanova, S.A., Tyukavina, A., Thau, D. et al., 2013. High-resolution global maps of 21st-century forest cover change. *Science* 342, 850–3. doi:10.1126/science.1244693

Hermosilla, T., Wulder, M.A., White, J.C., Coops, N.C., Hobart, G.W., 2015a. An integrated Landsat time series protocol for change detection and generation of annual gap-free surface reflectance composites. *Remote Sens. Environ.* 158, 220–234.

Hermosilla, T., Wulder, M.A., White, J.C., Coops, N.C., Hobart, G.W., Campbell, L.B., 2016. Mass data processing of time series Landsat imagery: Pixels to data products for forest monitoring. *Int. J. Digit. Earth* 9, 1035–1054. doi:10.1080/17538947.2016.1187673

Holden, C.E., Woodcock, C.E., 2016. An analysis of Landsat 7 and Landsat 8 underflight data and the implications for time series investigations. *Remote Sens. Environ.* 185, 16–36. doi:10.1016/j.rse.2016.02.052

Houghton, R.A., Goodale, C.L., 2004. Effects of Land Use Change on the Carbon Balance of Terrestrial Ecosystems. Ecosyst. L. Use Chang. Geophys. Monogr. Ser.

Huang, C., Goward, S., Masek, J., Gao, F., Vermote, E., Thomas, N., Schleeweis, K. et al., 2009a. Development of time series stacks of Landsat images for reconstructing forest disturbance history. *Int. J. Digit. Earth.* doi:10.1080/17538940902801614

Huang, C., Goward, S.N., Schleeweis, K., Thomas, N., Masek, J.G., Zhu, Z., 2009b. Dynamics of national forests assessed using the Landsat record: Case studies in eastern United States. *Remote Sens. Environ.* 113, 1430–1442. doi:10.1016/j.rse.2008.06.016

Huang, C., Ling, P.-Y., Zhu, Z. 2015. North Carolina's forest disturbance and timber production assessed using time series Landsat observations. *Int. J. Digit. Earth* 8(12), 947–949.

Huang, C., Schleeweis, K., Thomas, N., Goward, S.N. 2011. Forest dynamics within and around the Olympic National Park assessed using time series Landsat observations. In Y. Wang (Ed.). *Remote Sensing of Protected Lands* (pp. 71–93). London: Taylor & Francis.

Huang, C., Song, K., Kim, S., Townshend, J.R.G., Davis, P., Masek, J.G., Goward, S.N., 2008. Use of a dark object concept and support vector machines to automate forest cover change analysis. *Remote Sens. Environ.* 112, 970–985. doi:10.1016/j.rse.2007.07.023

Huang, C., Thomas, N., Goward, S.N., Masek, J.G., Zhu, Z., Townshend, J.R.G., Vogelmann, J.E., 2010. Automated masking of cloud and cloud shadow for forest change analysis using Landsat images. *Int. J. Remote Sens.* doi:10.1080/01431160903369642

Hurtt, G.C., Pacala, S.W., Moorcroft, P.R., Caspersen, J., Shevliakova, E., Houghton, R.A., Moore, B., 2002. Projecting the future of the U.S. carbon sink. *Proc. Natl. Acad. Sci. USA* 99, 1389–1394. doi:10.1073/pnas.012249999

Kennedy, R.E., Yang, Z., Cohen, W.B., 2010. Detecting trends in forest disturbance and recovery using yearly Landsat time series: 1. LandTrendr—Temporal segmentation algorithms. *Remote Sens. Environ.* 114, 2897–2910. doi:10.1016/j.rse.2010.07.008

Kennedy, R.E., Yang, Z., Braaten, J., Copass, C., Antonova, N., Jordan, C., Nelson, P., 2015. Attribution of disturbance change agent from Landsat time-series in support of habitat monitoring in the Puget Sound region, USA. *Remote Sens. Environ.* 166, 271–285.

King, A.W., Dilling, L., Zimmerman, G.P., Fairman, D.M., Houghton, R.A., Marland, G., Rose, A.Z., Wilbanks, T.J., 2007. *The first state of the carbon cycle report SOCCR: The North American carbon budget and implications for the global carbon cycle.* The First State of the Carbon Cycle Report SOCCR: The North American Carbon Budget and Implications for the Global Carbon Cycle 22, 242.

Masek, J.G., Goward, S.N., Kennedy, R.E., Cohen, W.B., Moisen, G.G., Schleeweis, K., Huang, C., 2013. United States forest disturbance trends observed using Landsat time series. *Ecosystems* 16, 1087–1104. doi:10.1007/s10021-013-9669-9

Masek, J.G., Vermote, E.F., Saleous, N.E., Wolfe, R., Hall, F.G., Huemmrich, K.F., Gao, F., Kutler, J., Lim, T.K., 2006. A landsat surface reflectance dataset for North America, 1990–2000. *IEEE Geosci. Remote Sens. Lett.* 3, 68–72. doi:10.1109/LGRS.2005.857030

McRoberts, R.E., Walters, B.F., 2012. Statistical inference for remote sensing-based estimates of net deforestation. *Remote Sens. Environ.* 124, 394–401. doi:10.1016/j.rse.2012.05.011

Michalak, A.M., Hirsch, A., Bruhwiler, L., Gurney, K.R., Peters, W., Tans, P.P., 2005. Maximum likelihood estimation of covariance parameters for Bayesian atmospheric trace gas surface flux inversions. *J. Geophys. Res. Atmos.* 110, 1–16. doi:10.1029/2005JD005970

Moisen, G.G., Meyer, M.C., Schroeder, T.A., Toney, C.J., Liao, X., Schleeweis, K., Freeman, E.A., 2016. Shape selection in Landsat time series: a tool for monitoring forest dynamics. *Glob. Chang. Biol.* 22(10), 3518–3528.

Neigh, C.S., Bolton, K.D., Diabate, M., Williams, J.J., Carvalhais, N., 2014. An automated approach to map the history of forest disturbance from insect mortality and harvest with Landsat time-series data. *Remote Sens.* 6(4), 2782–2808.

Nemani, R., Votava, P., Michaelis, A., Melton, F., and Milesi, C., 2011. Collaborative supercomputing for global change science. *Eos, Trans. Amer. Geophys. Union* 92, 109–110.

Olofsson, P., Foody, G.M., Herold, M., Stehman, S. V., Woodcock, C.E., Wulder, M.A., 2014. Good practices for estimating area and assessing accuracy of land change. *Remote Sens. Environ.* 148, 42–57.

Omerik, J.M., 1987. Ecoregions of the conterminous United States. *Ann. Assoc. Am. Geogr.* 77, 118–125.

Reams, G.A., Smith, W.D., Hansen, M.H., Bechtold, W.A., Roesch, F.A., Moisen, G.G., 2005. The forest inventory and analysis sampling frame. In Chapter 2 of *The Enhanced Forest Inventory and Analysis Program—National Sampling Design and Estimation Procedures.* W.A. Bechtoldand, P.L. Patterson Eds. USDA Forest Service, Asheville, NC. General Technical Report SRS-80 pp. 11–26.

Roy, D.P., Ju, J., Kline, K., Scaramuzza, P.L., Kovalskyy, V., Hansen, M., Loveland, T.R., Vermote, E., Zhang, C., 2010. Web-enabled Landsat Data WELD: Landsat ETM+ composited mosaics of the conterminous United States. *Remote Sens. Environ.* 114, 35–49. doi:10.1016/j.rse.2009.08.011

Ruefenacht, B., Finco, M.V., Nelson, M.D., Czaplewski, R., Helmer, E.H., Blackard, J.A., Holden, G.R. et al., 2008. Conterminous US and Alaska forest type mapping using forest inventory and analysis data. *Photogramm. Eng. Remote Sens.* 74, 1379–1388. doi:10.14358/PERS.74.11.1379

Schleeweis, K., Goward, S.N., Huang, C., Dwyer, J.L., Dungan, J.L., Lindsey, M.A., Michaelis, A., Rishmawi, K., Masek, J.G., 2016. Selection and quality assessment of Landsat data for the North American forest dynamics forest history maps of the US. *Int. J. Digit. Earth* 9, 963–980. doi:10.1080/17538947.2 016.1158876

Schroeder, T.A., Wulder, M.A., Healey, S.P., Moisen, G.G., 2011. Mapping Wildfire and Clearcut Harvest Disturbances in Boreal Forests with Landsat Time Series Data. *Remote Sens. Environ.* 115, 1421–1433.

Smith, W.B., 2002. Forest inventory and analysis: A national inventory and monitoring program. *Environ. Pollut.* 116, 233–242. doi:10.1016/S0269-7491(01)00255-X

Smith, W.B., Miles, P.D., Perry, C.H., Pugh, S.A., 2009. *Forest resources of the United States, 2007.* USDA Forest Service General Technical Report WO-78. Washington, DC: U.S. Department of Agriculture.

Stehman, S. V., 2013. Estimating area from an accuracy assessment error matrix. *Remote Sens. Environ.* 132, 202–211. doi:10.1016/j.rse.2013.01.016

Tans, P.P., Fung, I.Y., Takahashi, T., 1990. Observational constraints on the global atmospheric CO_2 budget. *Science* 247, 1431–1438.

Thomas, N.E., Huang, C., Goward, S.N., Powell, S., Rishmawi, K., Schleeweis, K., Hinds, A., 2011. Validation of North American forest disturbance dynamics derived from Landsat time series stacks. *Remote Sens. Environ.* 115, 19–32. doi:10.1016/j.rse.2010.07.009

Thornton, P.E., Law, B.E., Gholz, H.L., Clark, K.L., Falge, E., Ellsworth, D.S., Goldstein, A.H., Monson, R.K., Hollinger, D., Falk, M., Chen, J., Sparks, J.P., 2002. Modeling and measuring the effects of disturbance history and climate on carbon and water budgets in evergreen needleleaf forests. *Agric. For. Meteorol.* 113, 185–222. doi:10.1016/S0168-19230200108-9

Townshend, J.R., Masek, J.G., Huang, C., Vermote, E.F., Gao, F., Channan, S., Sexton, J.O., Feng, M., Narasimhan, R., Kim, D., Song, K., Song, D., Song, X.-P., Noojipady, P., Tan, B., Hansen, M.C., Li, M., & Wolfe, R.E., 2012. Global characterization and monitoring of forest cover using Landsat data: opportunities and challenges. *International Journal of Digital Earth* 5, 373–397.

Vogelmann, J.E., Howard, S.M., Yang, L., Larson, C.R., Wylie, B.K., Van Driel, J.N., 2001. Completion of the 1990s national land cover data set for the conterminous United States. *Photogramm. Eng. Remote Sens.* 67, 650–662.

Vogelmann, J.E., Xian, G., Homer, C., Tolk, B., 2012. Monitoring gradual ecosystem change using Landsat time series analyses: Case studies in selected forest and rangeland ecosystems. *Remote Sens. Environ.* 122, 92–105. doi:10.1016/j.rse.2011.06.027

Waring, R., Running S.W., 1998. *Forest Ecosystems: Analysis at Multiple Scales.* San Diego: Academic Press.

White, J.C., Wulder, M.A., Hobart, G.W., Luther, J.E., Hermosilla, T., Griffiths, P., Coops, N.C. et al., 2014. Pixel-based image compositing for large-area dense time series applications and science. *Can. J. Remote Sens.* 40, 192–212. doi:10.1080/07038992.2014.945827

Williams, C.A., Collatz, G.J., Masek, J., Huang, C., Goward, S.N., 2014. Impacts of disturbance history on forest carbon stocks and fluxes: Merging satellite disturbance mapping with forest inventory data in a carbon cycle model framework. *Remote Sens. Environ.* 151, 57–71.

Woodcock, C.E., Allen, R., Anderson, M., Belward, A., Bindschadler, R., Cohen, W., Gao, F. et al., 2008. Free access to Landsat imagery. *Science* 320, 1011–1012. doi:10.1126/science.320.5879.1011a

Wulder, M.A., Masek, J.G., Cohen, W.B., Loveland, T.R., Woodcock, C.E., 2012. Opening the archive: How free data has enabled the science and monitoring promise of Landsat. *Remote Sens. Environ.* 122, 2–10. doi:10.1016/j.rse.2012.01.010

Zhao, F., Huang, C., Zhu, Z., 2015. Use of vegetation change tracker and support vector machine to map disturbance types in greater yellowstone ecosystems in a 1984–2010 Landsat time series. *IEEE Geoscience. Remote Sens. Lett.* 12, 1650–1654.

Zhao, F.R., Meng, R., Huang, C., Zhao, M., Zhao, F.A., Gong, P., Yu, L., Zhu, Z., 2016. Long-term post-disturbance forest recovery in the greater Yellowstone ecosystem analyzed using Landsat time series stack. *Remote Sens.* 8, 898.

Zhu, Z., Woodcock, C.E., Holden, C., Yang, Z., 2015. Generating synthetic Landsat images based on all available Landsat data: Predicting Landsat surface reflectance at any given time. *Remote Sens. Environ.* 162, 67–83. doi:10.1016/j.rse.2015.02.009

10

Monitoring Annual Vegetated Land Loss to Urbanization with Landsat Archive: A Case Study in Shanghai, China

Qingling Zhang and Bhartendu Pandey

CONTENTS

10.1 Introduction..205
10.2 Methodology 207
    10.2.1 Data and Preprocessing 207
    10.2.2 Constructing Time Series of Annual
           Cloud/Shadow Free Landsat NDVI Composites 208
    10.2.3 Properties of the NDVI Mosaics 209
    10.2.4 Simulating NDVI Trajectory Models 209
    10.2.5 Pinpointing Changes..210
    10.2.6 Post Processing..211
    10.2.7 Accuracy Assessment and Evaluation..211
10.3 Results 212
    10.3.1 Results of Change Detection 212
    10.3.2 Change Detection Accuracy..212
    10.3.3 Comparison with Official Statistics Data 213
10.4 Discussions 214
10.5 Conclusions...217
10.6 Acknowledgment...217
References..217

10.1 Introduction

Due to industrialization, economic growth, technological advancement and population explosion, urban expansion (or urbanization hereafter) is the most irreversible and human-dominated form of land-use change, which modifies land cover, hydrological systems, biogeochemistry, climate, and biodiversity (Grimm et al., 2008). Urban expansion often causes significant disturbance to ecosystems surrounding cities, mostly resulting in the removal of large amounts of biomass and in turn putting the human-nature systems at risk.

Globally, urban expansion is one of the primary drivers of habitat loss and species extinction (Hahs et al., 2009). In many developing countries, urban expansion occurs on prime agricultural lands (Seto et al., 2000; del Mar López et al., 2001).

Such dynamics are especially significant in China. Since its opening and reforming in 1978, China's urbanized areas have almost tripled and this expansion is projected to continue in the future. The urban population in China was less than 20% in 1978, but rapidly increased to 52% in 2012 (United Nations, 2014), an increase of more than 500 million people. Consequently, cropland loss in China is significant due to the fact that urbanization has often been occurring over fertile agricultural land. This poses a great challenge to China's food security and has led to a big debate in and out of the country: who is going to feed the Chinese population? However, development pressure is also high that further increases urban land requirements and thereby competition with agricultural land-use. China depends on its cities for economic growth and innovation, and ultimately aims to concentrate about 70% of its population, about 900 million people, into cities by 2025 (Johnson, 2013). According to the World Bank, at 54%, China's degree of urbanization is still well below the 70% expected of a country with its current income level per person. The flood of migrants will continue, and by 2030 Chinese cities will contain more than 1 billion people (World Bank and Development Research Center of the State Council, the People's Republic of China, 2014).

High temporal frequency monitoring is critical to assessing land policy outcomes in addition to gaining an in-depth understanding of the size, type, and rate dynamics of urban areas. Studies that have moved beyond mapping to link social and economic processes to land use have shown that monitoring change for multiple periods (i.e., three or more every 5 years) is pivotal to understand the complex drivers of urban morphology through space and time, and to forecast future land use trends (Seto & Kaufmann, 2003). In China, the urbanization process is often driven by two major forces: development and cropland protection. On one hand, to feed its gigantic population, primary crop lands are strictly protected by the Chinese central government with strict land protection laws. On the other hand, to boost economy, local governments often experience high pressure to convert cropland to build houses, factories, and infrastructure. Such a pressure becomes of paramount importance to address when housing prices rise to a level that citizens cannot bear and price level ultimately becomes contentious. As its economy continues to grow, cities are continually expanding in China. To deal with such a situation, local governments often play a numbers game with the central government: report less cropland area loss in their official yearly statistical reports. In 2006, the central government passed a special development policy in Shanghai to relax cropland protection requirements in order to stimulate development in the Pudong district of Shanghai. A broader context at the time is that the housing market continued to boom from 2000 to 2006, but land provision for housing declined dramatically during the same period. The governments faced high pressure to provide more land for housing to cool down the housing market. The special land policy directly led to the dramatic increase in cropland area loss reported in the 2006 statistical yearbook.

Landsat imagery has long been utilized to monitor urbanization and ecosystem change at regional and local scales (Seto et al., 2011). However, fewer studies used Landsat time series to monitor urbanization at higher temporal frequencies, especially for large area applications, mainly due to the lack of efficient algorithms and computational facilities to handle large data volumes. Very recently, Schneider et al. (2012), Verbesselt et al. (2012), Hansen et al. (2013), Castrence et al. (2014), Xue et al. (2014), Hermosilla et al. (2015), Dutrieux et al. (2015), Zhu et al. (2016), Zhang & Weng (2016), White et al. (2017), and Zhu (2017) used Landsat time series to track annual or finer temporal resolution ecosystem changes. These efforts prove the effectiveness of Landsat time series in detecting high temporal resolution land-use and land-cover change.

Here we extract annual vegetated land conversion to urbanization information with Landsat time series and implement it on the Google Earth Engine (GEE) cloud-computing platform for large area applications. We first generate annual Landsat cloud/shadow free Normalized Difference Vegetation Index (NDVI) mosaics with a compositing method proposed by Zhang et al. (2015) and then construct NDVI time series spanning 2000–2010 with a rolling strategy. Changes due to the removal of large amounts of biomass can lead to a sudden drop in NDVI values, which can be well captured by the constructed Landsat NDVI time series at a 30 m resolution, considering the relatively small spatial scales of annual urban expansion.

10.2 Methodology

10.2.1 Data and Preprocessing

The Google Earth Engine gathers together the Earth's raw satellite imagery data petabytes of historical, present, and future data, including those from Landsat—and makes them easily available for analysis via expert provided algorithms (Moore & Hansen, 2011; Gorelick et al., 2017). The advanced computing system Google Earth Engine now allows efficiently processing and characterizing global-scale Landsat time series data sets for quantifying land change (Hansen et al., 2013). Google Earth Engine hosts the entire Landsat-7 Enhanced Thematic Mapper Plus (ETM+) data archive from the U.S. Geological Survey Earth Resources Observation and Science (EROS) Center Landsat archive. All the Landsat-7/ETM+ L1T data, regardless of the amount of cloud cover, are used in this study. Radiometrically calibrated and orthorectified using ground control points and digital elevation model (DEM) data to correct for relief displacement, the L1T geo-location error is less than 30 m even in areas with substantial terrain relief (Lee et al., 2004). These are the highest quality Level-1 products suitable for pixel-level time series analysis. Each Landsat-7/ETM+ L1T scene is approximately 185 km by 185 km and since May 2003 has had 22% missing pixels occurring in a repeating along-scan stripe pattern because of the ETM+ scan line corrector (SLC) failure (Markham et al., 2004). In this study, we use all the six 30 m reflective Landsat ETM+ wavelength bands: Blue (band 1: 0.45–0.52 μm), Green (band 2: 0.53–0.61 μm), Red (band 3: 0.63–0.69 μm), Near infrared (band 4: 0.78–0.90 μm), and two Middle infrared (bands 5 and 7: 1.55–1.75 μm and 2.09–2.35 μm). We use all the available Landsat-7/ETM+ imagery from 2000 to 2012 to generate nominal annual NDVI composites to form a time series for further analysis.

We use the Top of Atmosphere (TOA) reflectance for Landsat-7/ETM+, which was generated by using coefficients derived from Chander et al. (2009). Bad pixels with at least one spectral band missing often exist in ETM+ images, especially at the edges of a scene. These bad image pixels are an indicator of degraded data quality, and must be first identified and masked out before generating the NDVI composite. We generate a bad ETM+ pixel (including gaps due to SLC-off) mask by examining the product of Bands 1–5 and Band 7: if the product equals 0, the pixel is a bad one and must be discarded.

10.2.2 Constructing Time Series of Annual Cloud/Shadow Free Landsat NDVI Composites

We first generate annual cloud/shadow free NDVI composites from Landsat imagery time series, applying the mixed NDVI compositing strategy to build nominal annual NDVI mosaics (Zhang et al., 2015). In detecting urban areas with remote sensing imagery, fallow land often poses the greatest challenge, due to its spectral similarity to urban features, such as roads, roofs, and so on. To maximize the degree of separability between fallow lands and urban areas, we build a nominal annual composite with Landsat images taken within a contiguous three year period. This is based on the fact that on one hand, fallow lands are normally temporal. On the other hand, urbanization is often path-dependent, that is, a piece of land once built up is very unlikely to return back to other types of land within a relatively short period, for example, three years. Furthermore, taking images from three years can maximize the chance of getting cloud and cloud shadow free observations, especially in regions that have heavy cloud cover and after the ETM+ SLC failure. We build a nominal annual composite for one specific year with a time series comprising all imagery from that year and the following two years. In that way, consistent urban areas can be successfully separated from temporal fallow lands.

To build a time series of NDVI mosaics, we use a rolling compositing method. We first build the 2000 annual NDVI mosaic with ETM+ imagery in the 3 year period: 2000, 2001, and 2002, then build the following 2001 annual NDVI mosaic with the imagery in the following 3 years: 2001, 2002, and 2003, and continue with this method. With such a rolling compositing strategy, we are able to maximize the separability between existing cropland and urban areas

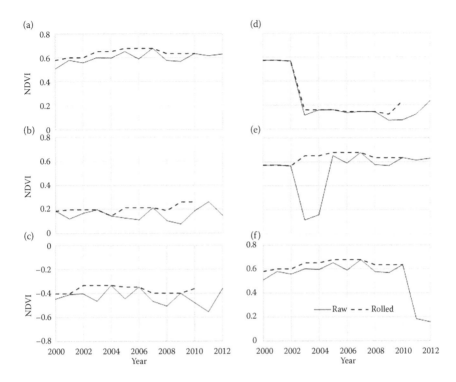

FIGURE 10.1
Effects of the rolling composing process on different trajectories. (a) stable cropland, (b) stable urban, (c) stable water body, (d) cropland lost since 2002, (e) temporal disturbance to cropland, (f) temporal cropland conversion to bare land at the end of the time series.

and preserve the trajectory of change at the same time. Figure 10.1 elucidates the effects of rolling on simulated stable and change NDVI trajectories. It is clear that after the rolling processes, stable signals remain stable (a, b, c), long-term change signal remains almost the same, and most importantly, the time point of change is well preserved (d). However, temporal changing signals are smoothed out (e, f), which is desired in the current study as discussed above.

10.2.3 Properties of the NDVI Mosaics

As shown by Zhang et al. (2015), the composed Landsat NDVI mosaics exhibit a tri-mode distribution of NDVI: water, bare land including urban, and vegetation falling into distinguished ranges (Figure 10.2). This indicates that vegetation conversion to urban signals will be easily captured with the composed Landsat NDVI time series.

10.2.4 Simulating NDVI Trajectory Models

The NDVI trajectory classes can be grouped into 11 categories: 10 change classes plus a stable class and a simulated theoretical model for each of them

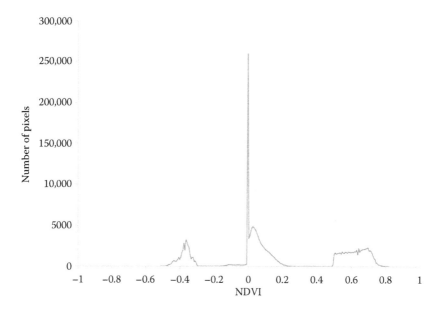

FIGURE 10.2
The distribution of NDVI values in the 2000 Landsat NDVI composite (regenerated with NDVI data in Shanghai based on Zhang, Q. et al. 2015. *Remote Sensing*, 2015(7), 11887–11913).

can be used as a cookie cutter for further identifying real observed NDVI trajectories. Bearing in mind that reconstructing the history of cropland conversion to urban use is the major focus of this study, the 10 change classes are defined with the year of change, as shown in Figure 10.3. For example, the 2000 class is defined as a pixel changed from vegetated land to urban land in year 2000 and its NDVI value dropped significantly in year 2001. Based on the tri-modal distribution of NDVI, we simulate stable NDVI with a mean of 0.6 and a standard deviation of 0.05, stable bare land with a mean of 0.1 and a standard deviation of 0.05. The inclusion of a standard deviation is to add some random noise into the simulated time series to make them more realistic. We further simulate a stable model for each given NDVI time series with its mean and a standard deviation of 0.05.

10.2.5 Pinpointing Changes

Given a pixel with a defined NDVI time series, we identify its shape by comparing it against the 11 simulated models with minimum distance criteria. To do that, we first calculate Euclidean distances between that given NDVI time series and each simulated model.

$$d_i = \sqrt{(x_1 - y_{1i})^2 + \cdots + (x_n - y_{ni})^2} \qquad (10.1)$$

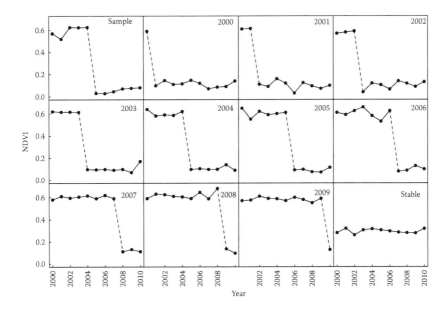

FIGURE 10.3
Models of NDVI trajectories.

where d_i is the Euclidean distance between a given NDVI time series and the simulated model i, x_n the NDVI value in the nth year, and y_{ni} is the simulated NDVI value of the simulated model i in the nth year.

If it is a stable time series, then the distance between it and the simulated stable model should be the minimum. For a change time series, the distance between it and its corresponding change model should be the minimum. Thus examining the minimum distance enables us to correctly label a given time series with an appropriate model, which then allows us to pinpoint the year of change if there is a change.

10.2.6 Post Processing

A built-up area normally has low positive NDVI. Thus for a change pixel from cropland to urban, its NDVI value at the final stage should be in the range of [0, 0.3]. Considering that shadows cast by tall buildings within urban centers can have negative NDVI values (Figure 10.2), we relax that range to [−0.2, 0.3] to represent all built-up areas. Thus any labeled change class with an NDVI value not in the range of [−0.2, 0.3] in 2010 should be relabeled. We add this check with the purpose of minimizing commission errors.

10.2.7 Accuracy Assessment and Evaluation

To assess the accuracy of our method to pinpoint time of change, we generate a set of random points using a stratified random sampling scheme so there are

30 points for each class (a total of 330 points) and have them analyzed by three analysts with cross checking until the class label of each point is agreed upon by all the three analysts to minimize human biases introduced in the ground reference data collection process. The three analysts did not participate in the change detection work and were only briefly instructed as to how the change and stable classes were defined. They rely on the composed NDVI time series and Google Earth's historic high resolution images to label the samples. There can be disagreement on some samples. In that case, the three analysts were asked to sit down together and go over those points to make a final decision about how to label them. There were about 20% of samples labeled through this process. Using these reference points, we calculate the detection accuracy.

We also collected cropland area data from statistics yearbooks of Shanghai for all the years and compared them with our results.

10.3 Results

10.3.1 Results of Change Detection

Figure 10.4 shows cropland to urban conversion history during the period from 2000 to 2010 detected with NDVI time series in Shanghai. The spatio-temporal dynamics are well captured: urban area expanded gradually from the existing urban cores out to the suburban areas from time to time. The northern part, the Chongming Island, experienced very little urbanization during this period. Pixels belonging to the same major roads were almost constructed in the same time (Figure 10.5), which is in good agreement to our common sense. In China, urbanization is often a top-down process, which is designed, implemented, and controlled by local governments through massive investments in infrastructures such as roads and land development policies.

10.3.2 Change Detection Accuracy

The proportion of the pixels that have the same change time between the results and the reference data is referred to as the "change detection accuracy" (Zhu & Woodcock, 2014). The proportion of the pixels that have the same time of change between the algorithm results and reference data is 84.87% (Table 10.1). The proportion of the pixels that are found to have changed earlier or later than the reference data but are within 1 year when a change is observable is 8.8% (Table 10.1). A change might have occurred in the second half of the previous year after the growing season, consequently the pixel's annual maximum NDVI value in that year did not change after going through

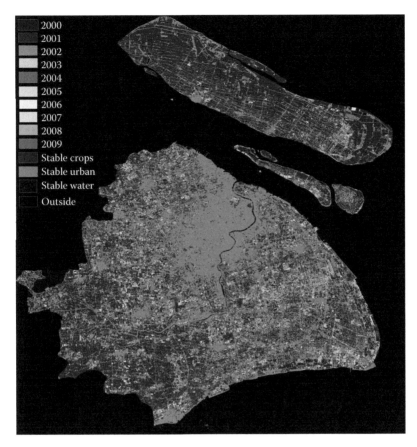

FIGURE 10.4
(**See color insert.**) Annual agriculture land loss in Shanghai from 2000 to 2010 (Dark green: stable agriculture; Dark blue: stable water; Gray: stable urban area; Black: out of Shanghai).

the 3-year composing process. Thus the algorithm will report the change as occurring in the following year (1 year later). Similarly, a pixel might have been cleaned for construction, but buildings were not built in the current year or even in the following few years. Due to limited number of high resolution images available in Google Earth, analysts might have reported a later change. This is the reason that a later change error more than 1 year is very rare while there are some earlier change errors more than 1 year (Table 10.1).

10.3.3 Comparison with Official Statistics Data

The patterns from both datasets are generally compatible: there is a peak in year 2006 (Figure 10.6). What needs to be pointed out is that the statistic yearbook reports are consistently lower than the estimation from satellite, except for years 2003 and 2006.

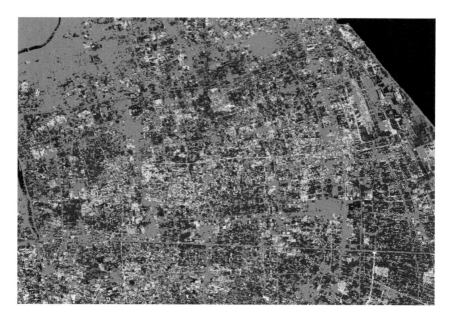

FIGURE 10.5
(**See color insert.**) Annual agriculture land loss in Shanghai from 2000 to 2010 in part of the Pudong district, Shanghai, China. Colors indicate times of conversion (Gray: stable land cover).

10.4 Discussions

The success of the "cookie cutter" method relies on the ETM+ NDVI mosaics. In relatively coarse resolution MODIS NDVI data, gradual change is very common. Whereas, with the 30 m spatial resolution of Landsat imageries, small-scale changes are detectable. In fact, in an urbanizing region, a majority of changes are abrupt in nature, which can be conveniently captured by the "cookie cutter" method.

Urbanization is often a very dynamic process, changing from year to year. Such a highly dynamic process is well demonstrated in Shanghai with satellite observations and data from official statistical reports, indicating the need for high frequency observations in order to understand the process and the underlying land use policies.

Observations from satellites capture the spatial pattern of urban expansion, which is lacking in official statistical reports. The success of our method to capture spatio-temporal dynamics of ecosystem disturbances caused by urbanization is mainly attributed to the 30 m spatial resolution and 16-day temporal resolution of the Landsat data. The free access to the entire Landsat archive and the availability of the GEE cloud computing facility makes it possible for us to develop efficient algorithms to monitor such processes effectively and with high accuracies. In our study, we also noticed that change detection accuracies vary

TABLE 10.1

Accuracy of Cropland to Urban Land Change Detection

Reference	Predicted											Producer's Accuracy
	2001	2002	2003	2004	2005	2006	2007	2008	2009	2010	Stable	
2001	24											100.00
2002	3	28	3									82.35
2003		1	24	1								92.31
2004			2	23								92.00
2005				6	26	1					1	76.47
2006					1	24	2		1		1	82.76
2007						3	23	1	2			79.31
2008							1	26	1		1	89.66
2009								2	23	1		88.46
2010										24	1	96.00
Stable	3	1	1		3	2	3	2	3	5	86	78.90
User's Accuracy	80.00	93.33	80.00	76.67	86.67	80.00	76.67	86.67	76.67	80.00	95.56	
Overall Accuracy (%)				84.87								
							K-hat Statistic (%)		82.81			

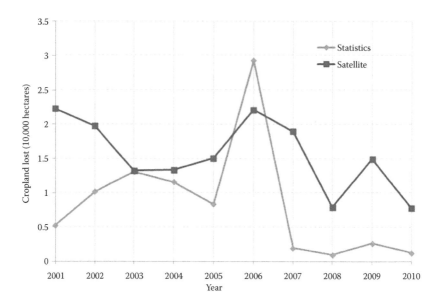

FIGURE 10.6
Area of annual cropland loss in Shanghai estimated from satellite data and statistic yearbooks.

from year to year. One possible reason could be variable Landsat data quality. As discussed before, the quality of ETM+ imagery can be degraded by many factors, such as missing scan lines and the existence of clouds/shadows, although we tried to alleviate these problems through a 3-year compositing strategy. In the current study, we used only the Landsat-7/ETM+ data. Now with the availability of Landsat-5/Thematic Mapper (TM), Landsat-8/Operational Land Imager (OLI), and Sentinel-2A/B data, we can consider merging Landsat-7/ETM+ data with them to improve change detection accuracy.

Compared with other methods, such as the widely implemented post classification change detection method in Zhu et al. (2016) as well as in Zhang & Weng (2016), our cookie cutter method can yield a high change detection accuracy and at the same time is simple, automated, and easy to implement. On the one hand, unlike supervised classification schemes, our "cookie cutter" method does not require collection of training samples, which is often a burden especially for applications over very large regions. On the other hand, unlike unsupervised classification schemes, our "cookie cutter" method does not require labeling classes after the classification. Labeling each pixel is automatically done during the classification process, without the need for human intervention. Thus, its per-pixel nature allows it to be directly applied to a larger region on the GEE platform. Yet, its performance over larger geographical settings needs to be further investigated.

Finally, as we only use NDVI time series to detect cropland loss, we are not able to capture cropland conversion to urban green spaces where NDVI values often quickly recover after disturbance.

10.5 Conclusions

We built annual NDVI mosaics from 2000 to 2010 by applying a rolling 3-year image compositing strategy. The compositing strategy uses a stratified NDVI mosaicking method applied to a dense Landsat ETM+ time stacks collected in three contiguous years.

We simulate 11 change models and a stable model as "cookie cutters" to reconstruct cropland to urban conversion history. Minimum Euclidean distance between a given Landsat NDVI time series and the 12 simulated models is used to pinpoint the timing of change or label the pixel as temporally stable.

Our method is simple, easy, fast, and straightforward to understand. It forms a solid base for accurately monitoring ecosystem disturbance due to urban expansion in a large region, thanks to the fine spatial and temporal resolutions of Landsat imagery and the GEE cloud computing facility.

Lastly, annual cropland loss estimation at a 30 m resolution will greatly enable our understanding of urban development trajectories, food security, and sustainable development.

10.6 Acknowledgment

This work was supported in part by the National Key Research and Development Program of China (2017YFB0504204) and the One Hundred Talents Program of the Chinese Academy of Science (Zhang et al., 2015).

References

Castrence, M., Nong, D., Tran, C., Young, L., & Fox, J. 2014. Mapping urban transitions using multi-temporal Landsat and dmsp-ols night-time lights imagery of the red river delta in Vietnam. *Land*, 3(1), 148–166.

Chander, G., Markham, B.L., & Helder, D.L. 2009. Summary of current radiometric calibration coefficients for Landsat MSS, TM, ETM+, and EO-1 ALI sensors. *Remote Sensing of Environ*, 113, 893–903.

del Mar López, T., Aide, T.M., & Thomlinson, J.R. 2001. Urban expansion and the loss of prime agricultural lands in Puerto Rico. *AMBIO: A Journal of the Human Environment*, 30, 49–54.

Dutrieux, L. P., Verbesselt, J., Kooistra, L., & Herold, M. 2015. Monitoring forest cover loss using multiple data streams, a case study of a tropical dry forest in Bolivia. *Isprs Journal of Photogrammetry & Remote Sensing*, 107, 112–125.

Gorelick, N., Hancher, M., Dixon, M., Ilyushchenko, S., Thau, D., & Moore, R. 2017. Google earth engine: Planetary-scale geospatial analysis for everyone. *Remote Sensing of Environment*, 202, 18–27.

Grimm, N.B., Faeth, S.H., Golubiewski, N.E., Redman, C.L., Wu, J., Bai, X., & Briggs, J.M. 2008. Global change and the ecology of cities. *Science*, 319, 756–760.

Hahs, A., McDonnell, M., McCarthy, M. et al. 2009. A global synthesis of plant extinction rates in urban areas. *Ecology Letters*, 12, 1165–1173.

Hansen, M.C., Potapov, P.V., Moore, R. et al. 2013. High-resolution global maps of 21-st century forest cover change. *Science*, 342, 850–853.

Hermosilla, T., Wulder, M.A., White, J.C., Coops, N.C., & Hobart, G.W. 2015. Regional detection, characterization, and attribution of annual forest change from 1984 to 2012 using Landsat-derived time-series metrics. *Remote Sensing of Environment*, 170, 121–132.

Johnson, I. 2013. China's great uprooting: Moving 250 million into cities. *New York Times*.

Lee, D.S., Storey, J.C., Choate, M.J., & Hayes, R.W. 2004. Four years of Landsat-7 on-orbit geometric calibration and performance. *IEEE Transactions on Geoscience and Remote Sensing*, 42, 2786–2795.

Markham, B.L., Storey, J.C., Williams, D.L., & Irons, J.R. 2004. Landsat sensor performance: History and current status. *IEEE Transactions on Geoscience and Remote Sensing*, 42, 2691–2694.

Moore, R.T., & Hansen, M.C. 2011. Google Earth Engine: A new cloud-computing platform for global-scale earth observation data and analysis. *AGU Fall Meeting Abstracts*. San Francisco, CA.

Schneider, A. 2012. Monitoring land cover change in urban and peri-urban areas using dense time stacks of Landsat satellite data and a data mining approach. *Remote Sensing of Environment*, 124, 689–704.

Seto, K.C., Fragkias, M., Güneralp, B., & Reilly, M.K. 2011. A meta-analysis of global urban land expansion. *PloS One*, 6, e23777.

Seto, K.C., & Kaufmann, R.K. 2003. Modeling the drivers of urban land use change in the Pearl River Delta, China: Integrating remote sensing with socioeconomic data. *Land Economics*, 79, 106–121.

Seto, K.C., Kaufmann, R.K., & Woodcock, C.E. 2000. Landsat reveals China's farmland reserves, but they're vanishing fast. *Nature*, 406, 121.

United Nations. 2014. *World Urbanization Prospects: The 2012 Revision*. New York: United Nations.

Verbesselt, J., Zeileis, A., & Herold, M. 2012. Near real-time disturbance detection using satellite image time series. *Remote Sensing of Environment*, 123(123), 98–108.

White, J.C., Wulder, M.A., Hermosilla, T., Coops, N.C., & Hobart, G.W. 2017. A nationwide annual characterization of 25 years of forest disturbance and recovery for canada using Landsat time series. *Remote Sensing of Environment*, 194, 303–321.

World Bank and Development Research Center of the State Council, the People's Republic of China. 2014. *Urban China: Toward Efficient, Inclusive, and Sustainable Urbanization*. Washington, DC: World Bank. © World Bank. https://openknowledge.worldbank.org/handle/10986/18865 License: CC BY 3.0 IGO.

Xue, X., Liu, H., Mu, X., & Liu, J. 2014. Trajectory-based detection of urban expansion using Landsat time series. *International Journal of Remote Sensing*, 35(4), 1450–1465.

Zhang, L., & Weng, Q. 2016. Annual dynamics of impervious surface in the pearl river delta, china, from 1988 to 2013, using time series Landsat imagery. *ISPRS Journal of Photogrammetry & Remote Sensing*, 113(3), 86–96.

Zhang, Q., Li, B., Thau, D. et al. 2015. Building a better urban picture: Combining day and night remote sensing imagery. *Remote Sensing*, 2015(7), 11887–11913.

Zhu, Z. 2017. Change detection using Landsat time series: A review of frequencies, preprocessing, algorithms, and applications. *ISPRS Journal of Photogrammetry & Remote Sensing*, 130, 370–384.

Zhu, Z., Fu, Y., Woodcock, C.E. et al. 2016. Including land cover change in analysis of greenness trends using all available Landsat 5, 7, and 8 images: A case study from Guangzhou, china (2000–2014). *Remote Sensing of Environment*, 185, 243–257.

Zhu, Z., & Woodcock, C.E. 2014. Continuous change detection and classification of land cover using all available Landsat data. *Remote Sensing of Environment*, 144(1), 152–171.

Index

A

AARD, *see* Average absolute relative difference
Aboveground net primary productivity (ANPP), 75
ACCA, *see* Automated Cloud Cover Assessment
Accuracy assessment, 211–212
ACPs, *see* Annual cycle parameters
AD, *see* Average difference
Adaptive classification accuracy assessment, 146–148
Adaptive time series
  and full length time series, 144, 149–150
  generation, 142–143
ADR, *see* Annual disturbance rate
Advanced HOT (AHOT), 18
Advanced Very High Resolution Radiometer (AVHRR), 4, 79
Agricultural mapping and monitoring of crops, 74
Agriculture land loss, 213, 214
AHOT, *see* Advanced HOT
Ancillary data selection for contaminated pixel, 29–30
Annual cycle parameters (ACPs), 91, 94, 96, 115; *see also* Land surface temperature
  ACP3, 94–95
  ACP5, 95, 105
  ACP5 R^2 and ACP3 R^2, 107
  applications, 107
  Aqua daytime MODIS LST pattern and SUHI of Paris, 108
  calculation principle, 96
  for Central Europe, 96–98
  climatological SUHI analysis, 107–112
  collection-5 vs. collection-6, 98–102
  conducted experiments and data processed for this study, 92
  data and methods, 92
  as disaggregation kernels, 112–115

  DLST RMSE improvement, 114
  global radiation pattern, 105
  latitudinal changes of ACP3 values, 103
  latitudinal gradients in, 102–107
  mean annual SUHI, 109
  MODIS land cover and urban areas, 94–96
  MYD11A1 and MOD11A1 land surface temperatures, 92–94
  projection of MODIS land products, 93
  simulated global radiation data, 106
  SUHII for overpass times, 110
  Terra MODIS daytime LST observations, 95
Annual disturbance rate (ADR), 186
Annual land cover mapping limitations, 138
Annual LST cycle, 91
ANPP, *see* Aboveground net primary productivity
AROP, *see* Automated registration and orthorectification package
ASTER (Advanced Spaceborne Thermal Emission and Reflection Radiometer), 91
Automated Cloud Cover Assessment (ACCA), 8, 176
Automated registration and orthorectification package (AROP), 47
Average absolute difference (AAD), 51
Average absolute relative difference (AARD), 51
Average difference (AD), 51, 60
AVHRR, *see* Advanced Very High Resolution Radiometer

B

BCI, *see* Biophysical Composition Index
Bidirectional reflectance distribution function (BRDF), 138

Biophysical Composition Index (BCI),
         122, 125; *see also* Impervious
         surface estimation
BRDF, *see* Bidirectional reflectance
         distribution function
Brightness Temperature (BT), 13
BT, *see* Brightness Temperature

C

C5.0, 14
CCDC, *see* Continuous Change
         Detection and Classification
CDR, *see* Climate Data Record
CFmask, 17
Change detection
    accuracy, 212–213
    and classification of land cover, 81
    methods, 157
Change metrics, 157–158
Class commission errors, 184
Clear view compositing (CVC), 177
    algorithm, 178
Climate Data Record (CDR), 11
Climatological SUHI analysis, 107–112
Closest spectral fit method (CSF
         method), 26
Cloud, 17
    contamination, 177
    cover percentage calculation, 6
Cloud and cloud shadow detection, 4,
         17, 19; *see also* Landsat time
         series analysis
    accuracy, 17
    automated, 4
    based on multitemporal Landsat
         images, 14–16
    based on single-date Landsat image, 8
    challenges, 17–18
    characteristics of algorithms, 9–10
    cloud-height-estimation method, 13
    comparison of algorithms, 17
    future development, 18
    geometry-based, 12, 13
    haze/thin cloud removal, 18–19
    machine-learning-based, 14
    physical-rules-based, 8–9, 11–13
    shape-similarity-match approach, 12
    spatial information, 18

temporal frequency, 18
    water vapor absorption band, 8
CMLP, *see* Contextual multiple linear
         prediction
Coleambally Irrigation Area, 49
Collection 5 MLCT product, 138
Conterminous United States
         (CONUS), 175
Contextual multiple linear prediction
         (CMLP), 26
Continuous Change Detection and
         Classification (CCDC), 81
CONUS, *see* Conterminous United States
Cookie cutter method, 214, 216
Cropland
    loss in Shanghai, 216
    to urban land change detection
         accuracy, 215
CSF method, *see* Closest spectral fit
         method
CVC, *see* Clear view compositing

D

Dark, dense forest (DDF), 178
Data fusion method
    accuracy assessment, 60
    quantitative assessment, 61
DDF, *see* Dark, dense forest
Defense Meteorological Satellite
         Program's Operational Linescan
         System (DMSP/OLS), 121
DEMs, *see* Digital Elevation Models
Dictionary-pair learning based
         methods, 45
Different days of year (DOY), 71
Digital Elevation Models (DEMs), 11,
         207; *see also* Cloud and cloud
         shadow detection
Disaggregation kernels, 112
Disturbance map, 165–166; *see also* Robust
         reference dataset creation
    class, 180
Diurnal temperature cycle (DTC), 114
DLST, *see* Downscaled LST
DMSP/OLS, *see* Defense Meteorological
         Satellite Program's Operational
         Linescan System
Downscaled LST (DLST), 112

DOY, *see* Different days of year
DTC, *see* Diurnal temperature cycle
DTW, *see* Dynamic time warping
Dynamics of impervious surfaces, 127–128
Dynamic time warping (DTW), 123, 126

E

Earth Observation Satellite Company (EOSAT), 177
Earth Observing System Data and Information System (EOSDIS), 92
Earth Resources Observation and Science (EROS), 4, 11, 175, 176, 207
Enhanced STARFM (ESTARFM), 45
Enhanced Thematic Mapper Plus (ETM+), 5, 26, 124, 140, 165, 176, 207
Enhanced vegetation index (EVI), 36, 138, 140, 71, 79
EOSAT, *see* Earth Observation Satellite Company
EOSDIS, *see* Earth Observing System Data and Information System
ePCA, *see* Extended PCA
EROS, *see* Earth Resources Observation and Science
ESA, *see* European Space Agency's
ESTARFM, *see* Enhanced STARFM
ETM+, *see* Enhanced Thematic Mapper Plus
European Space Agency's (ESA), 64
EVI, *see* Enhanced vegetation index
Extended PCA (ePCA), 77

F

Fine-resolution time series, 44
  predicted, 53
Flexible spatiotemporal data fusion method (FSDAF), 46, 54–58, 62–64, 125; *see also* Spatiotemporal data fusion
  assessment of, 60–62
  flowchart of, 54
  MODIS and Landsat images, 59
  quantitative assessment of, 61

  scatter plots of actual and predicted values, 62
  simulated coarse and fine images, 59
  test experiment, 58–60
Fmask (Function of mask), 5, 11
Forest
  carbon content, 173
  disturbance events, 174
  mapping and ecosystem analyses, 74–76
Forest-disturbance mapping project, 173, 199–200; *see also* NAFD-NEX product generation; Vegetation Change Tracker
  adjustment of MMU, 189–191
  annual disturbance map, 188, 192
  clear view compositing algorithm, 178
  cloud contamination, 177
  CVC algorithm, 178
  image compositing, 177–179
  image preprocessing, 177
  image selection, 176–177
  map re-projection, 191–192
  MMU filtering algorithm, 191
  MMU Size for NAFD-NEX Product, 189
  NAFD CONUS grid, 192
  NAFD-NEX processing flow, 175–176
  p27/r29, 188
  pixel-based approaches, 199
  post processing, 184–185
  quality assessment, 185–189
  stack level QA metrics, 185
  WRS-2 path/row, 187, 190
Forest Inventory and Analysis (FIA), 174, 182
FSDAF, *see* Flexible spatiotemporal data fusion method

G

GCTP, *see* General Cartographic Transformation Package
GDP, *see* Gross Domestic Product
GEE, *see* Google Earth Engine
General Cartographic Transformation Package (GCTP), 192

Geometry-based cloud shadow
        detection approach, 12; *see also*
        Cloud and cloud shadow
        detection
Global radiation
    simulated data, 106
    temporal pattern of, 105
Google Earth Engine (GEE), 207
    cloud computing platform, 84
Greenness trend analyses, 78
Gross Domestic Product (GDP), 123

H

Haze/thin cloud removal, 18–19
HDF, *see* Hierarchical data format
HF, *see* Homomorphic Filter
Hierarchical data format (HDF), 93
High quality Landsat time series
        reconstruction, 26, 35, 38;
        see also Landsat time series
        analysis
    ancillary data selection for
        contaminated pixel, 29–30
    automatic system for, 28
    contaminated pixel interpolation,
        30–33
    experiments, 33
    k-means method, 29
    Landsat-7 images in Mona Island
        Site, 33, 34, 35
    methods, 28–29
    real, 35–36
    RMSE and R Values, 37
    similar pixel selection, 30
    simulated, 36–38
    spatial interpolators, 26
    spatiotemporal interpolators, 27
    temporal interpolators, 26–27
    uncontaminated pixels
        classification, 29
High resolution (HR), 122
High temporal frequency
        monitoring, 206
Homomorphic Filter (HF), 19
HR, *see* High resolution
Hydro-phenological analyses of
        complex flooded landscapes,
        76–77

I

IBRA, *see* Interim Biogeographic
        Regionalisation for Australia
IFOV, *see* Instantaneous field-of-view
IFZ, *see* Integrated forest z-score
IGBP, *see* International Geosphere
        Biosphere Programme
IHOT, *see* Iterative Haze Optimized
        Transformation
Image
    contamination, 25–26
    data, 140
    preprocessing, 177
    selection, 176–177
Image compositing, 177–179
    map re-projection, 191–192
    post processing, 184–185
    quality assessment, 185–189
Impervious surface, 121
    classification accuracy, 128–131
    classification based on semi-
        supervised SVM, 126–127
    dynamics of, 127–128
    procedures for mapping annual
        dynamics of, 125
    in Wuhan city, 123, 124, 129, 130
Impervious surface estimation, 121,
        127, 131
    data preprocessing, 124–125
    methodology, 123
    object-based analysis methods, 122
    per-pixel analysis methods, 122
    reconstruction of time series BCI,
        125–126, 128
    similarity of temporal features, 126
    study area, 123
    sub-pixel analysis methods, 122
Instantaneous field-of-view (IFOV), 90
Integrated forest z-score (IFZ), 81, 180
Interim Biogeographic Regionalisation
        for Australia (IBRA), 158
International Geosphere Biosphere
        Programme (IGBP), 94
ISODATA, *see* Iterative Self-Organizing
        Data Analysis Technique
Iterative Haze Optimized
        Transformation (IHOT), 15
Iterative Self-Organizing Data Analysis
        Technique (ISODATA), 47

J

Joint Polar Satellite System (JPSS), 64
JPSS, *see* Joint Polar Satellite System

K

K-means method, 29

L

L1T, *see* Level-1 terrain corrected
L7 Irish masks, 7
L8 SPARCS masks, 7
L8SR, *see* Landsat 8 surface reflectance
Land cover, 137
    change detection and classification
        of, 81
    classification, 140
    evaluation of trajectory mapping, 149
    impacts of changes in, 138
    mapping accuracy, 150
    mapping limitations, 138
    maps in June of each year, 147
    trajectories, 146
Land Processes Distributed Active
        Archive Center (LP DAAC), 92
Landsat-7 images in Mona Island Site,
        33, 34, 35
    false-color composite of, 34
Landsat 8 surface reflectance (L8SR), 124
Landsat CDR, *see* Landsat Surface
        Reflectance Climate Data
        Record
Landsat cloud and cloud shadow
        masks, 7
Landsat data, 5–6
Landsat Ecosystem Disturbance
        Adaptive Processing System
        (LEDAPS), 11, 124, 165
Landsat Global Archive Consolidation
        (LGAC), 4
Landsat images, 4, 25, 207
    contamination, 25–26
    sun/cloud/shadow geometry in, 13
    use of multitemporal, 17
Landsat satellites, 4, 5
Landsat-specific phenology algorithm
        (LPA), 82

Landsat Surface Reflectance Climate
        Data Record (Landsat CDR), 123
Landsat Time Series (LTS), 4
    need for, 180–182
    reconstruction, 35–36
    simulated, 36–38
Landsat time series analysis, 4, 19;
        see also Cloud and cloud
        shadow detection; High
        quality Landsat time series
        reconstruction
    global cloud cover percentage
        calculation, 6
    L7 Irish masks, 7
    L8 Biome masks, 7
    L8 SPARCS masks, 7
    Landsat 1–5 MSS bands, 6
    Landsat 4–5 TM bands, 6
    Landsat 7 ETM+ Bands, 6
    Landsat 8 OLI/TIRS Bands, 6
    Landsat data, 5–6
    masks of Landsat cloud and cloud
        shadow, 7
Landsat Times Series Stacks (LTSS), 180
Land surface emissivity (LSE), 90; *see
        also* Annual cycle parameters
    retrieval methods, 90–91
Land surface temperature (LST), 89,
        138; *see also* Annual cycle
        parameters
    annual LST cycle, 91
    data of MYD11A1 and MOD11A1, 93
    disaggregation kernels, 112
    downscaling, 112
    retrieval, 90
    retrieval algorithms, 90
    satellite sensor, 90
    Terra MODIS daytime LST
        observations, 95
    thermodynamic temperature, 91
Land surface water index (LSWI), 71
LCZ, *see* Local Climate Zones
LEDAPS, *see* Landsat Ecosystem
        Disturbance Adaptive
        Processing System
Level-1 terrain corrected (L1T), 165
LFC, *see* Low forest cover
LGAC, *see* Landsat Global Archive
        Consolidation

Local Climate Zones (LCZ), 111
Low forest cover (LFC), 189
LPA, *see* Landsat-specific phenology
    algorithm
LP DAAC, *see* Land Processes Distributed
    Active Archive Center
LSE, *see* Land surface emissivity
LST, *see* Land surface temperature
LSWI, *see* Land surface water index
LTK, *see* Luo Trishchenko Khlopenkov
LTS, *see* Landsat Time Series
LTSS, *see* Landsat Times Series Stacks
Luo Trishchenko Khlopenkov (LTK), 11

M

Mapping forest-disturbance changes,
    174, 199–200; *see also* Forest-
    disturbance mapping project
Mapping land cover trajectories, 137,
    145–146, 150–153
  accuracy assessment, 144–148
  adaptive and full length time series
      comparison, 144, 149–150
  algorithm, 141
  Collection 5 MLCT product, 138
  detecting change dates, 141–142
  evaluation of, 149
  generating adaptive time series,
      142–143
  image data, 140
  input features used for land cover
      classification, 140
  integrated training and classification,
      143–144
  land cover mapping accuracy, 150
  land cover maps in June of each
      year, 147
  limitations of, 138
  methods, 141
  modified SVM classification, 143
  reconstruction, 144
  reference data, 140–141
  study area, 139–140
  sub-annual change detection, 142
  from 2001 to 2010, 146
  workflow of, 141
MAST, *see* Mean Annual Surface
    temperature

Mean Annual Surface temperature
    (MAST), 91
MFmask, *see* Mountainous Fmask
Minimum Mapping Unit (MMU), 189
  filtering algorithm, 191
MLCT, *see* MODIS Land Cover Type
MLE, *see* Multiple Lines of Evidence
MMT, *see* Multisensor Multiresolution
    Techniques
MMU, *see* Minimum Mapping Unit
MNSPI, *see* Modified Neighborhood
    Similar Pixel Interpolator
Moderate Resolution Imaging
    Spectroradiometer (MODIS), 4,
    40, 43–44, 138
  land cover and urban areas, 94–96
  and Landsat images, 59
  LST pattern and SUHI of Paris, 108
Moderate resolution satellites, 4
Modified Neighborhood Similar Pixel
    Interpolator (MNSPI), 27
MODIS, *see* Moderate Resolution
    Imaging Spectroradiometer
MODIS Land Cover Type (MLCT), 138
Moisture-sensitive indices, 71
Mountainous Fmask (MFmask), 11
MSI, *see* MultiSpectral Instrument
MSS, *see* Multispectral Scanner
    System
Multiple Lines of Evidence (MLE), 160
Multisensor Multiresolution Techniques
    (MMT), 45
MultiSpectral Instrument (MSI), 64
Multispectral Scanner System (MSS),
    5, 199
Multitemporal Landsat images, 17, 19
Multi-year phenological inference, 77;
      see also Phenological inference
  continuous change detection and
      classification of land cover, 81
  curves, 81–82
  greenness trend analyses, 78
  percent above threshold, 82
  phenocam networks, 78
  phenocam observations, 77–78
  phenological analyses with seasonal
      data, 81
  remote sensing-based phenology
      applications, 81

trajectory-based landscape change
analyses, 79–81
trend analyses, 78–79
Multi-year phenological inference, 81–82

N

NACP, *see* North American Carbon
Program
Nadir BRDF-adjusted reflectance
(NBAR), 140
NAFD, *see* North American Forest
Dynamics
NAFD CONUS grid, 192
NAFD Image Selection and Processing
(NISP), 176
NAFD-NEX product generation, 192;
see also Forest-disturbance
mapping project
annual disturbance map class
definitions for national
mosaic, 193
class definitions for time integrated
disturbance map mosaic, 194
error matrix for overall map
classes, 197
error matrix in proportion of
area, 198
NAFD-NEX forest disturbance
map, 193
NAFD-NEX last integrated forest
disturbance map, 194
NAFD-NEX processing flow, 175–176
NAFD-NEX product at ORNL,
192–194
producer's and user's accuracy for
disturbance class with 95%
confidence envelope, 196
validation, 195–198
NASA Earth Exchange (NEX), 177
National Centers for Environmental
Prediction (NCEP), 11
National Dynamic Land Cover Dataset
(NDLCD), 79
National Land Cover Database
(NLCD), 191
1992 dataset, 187, 189
NBAR, *see* Nadir BRDF-adjusted
reflectance

NBR, *see* Normalized Burn Ratio
index
NCEP, *see* National Centers for
Environmental Prediction
NCSA, *see* Number of clear sky LST
acquisitions
NDLCD, *see* National Dynamic Land
Cover Dataset
NDSI, *see* Normalized Difference Snow
Index
NDVI, *see* Normalized Difference
Vegetation Index
NDVI linear mixing growth model
(NDVI-LMGM), 46–49, 62–64;
see also Spatiotemporal data
fusion
accuracy assessment for irrigated
area, 52
assessment for long term prediction,
53–54
assessment over spatial and temporal
contrasting regions, 51–53
Coleambally Irrigation Area, 49
comparison for prediction
accuracy, 51
flowchart of, 48
predicted fine-resolution time
series, 53
using spatial moving window, 49
test experiment, 49–50
NDVI-LMGM, *see* NDVI linear mixing
growth model
NDWI, *see* Normalized difference water
index
Near InfraRed (NIR), 5, 36
Neighborhood Similar Pixel
Interpolator (NSPI), 27, 47;
see also High quality Landsat
time series reconstruction
NEX, *see* NASA Earth Exchange
NIR, *see* Near InfraRed
NISP, *see* NAFD Image Selection and
Processing
NLCD, *see* National Land Cover
Database
Normalized Burn Ratio index
(NBR), 165
Normalized Difference Snow Index
(NDSI), 8

Normalized Difference Vegetation
      Index (NDVI), 8, 44, 71, 90,
      141, 161, 182, 207; *see also*
      Urbanization, monitoring
   cloud/shadow free composites,
      208–209
   contrast in mixed deciduous and
      conifer forest area, 75
   dynamics of regional average
      NDVI, 50
   finding year of change, 210–211
   mosaics, 209
   trajectories, 211
   trajectory simulation, 209–210
Normalized difference water index
      (NDWI), 71, 126
North American Carbon Program
      (NACP), 174, 192
North American Forest Dynamics
      (NAFD), 174, 200
NSPI, *see* Neighborhood Similar Pixel
      Interpolator
Number of clear sky LST acquisitions
      (NCSA), 91, 95

O

Oak Ridge National Laboratory
      (ORNL), 192
Oak Ridge National Laboratory
      Distributed Active Archive
      Center (ORNL DAAC), 174
Object-based analysis methods, 122
OLI, *see* Operational Land Imager
Operational Land Imager (OLI), 5,
      123, 216
ORNL, *see* Oak Ridge National
      Laboratory
ORNL DAAC, *see* Oak Ridge National
      Laboratory Distributed Active
      Archive Center

P

PAT, *see* Percent above threshold
PCA, *see* Principal component analysis
Percent above threshold (PAT), 82
Per-pixel analysis methods, 122
Phenocam, 77

greenness index, 78
   networks, 78
   observations, 77
Phenological analyses, 81
Phenological inference, 69; *see also*
      Multi-year phenological
      inference; Single-season
      phenological analyses; Single-
      year phenology
   applications and importance of
      ancillary factors, 82–84
   spectral vegetation indices used in, 71
   time series of remote sensing data, 70
Phenology, 69
   profiles of land cover types, 73
   trajectory for one growing season, 72
   variation represented by seasonal
      trajectories, 72–73
Pixel-based approaches, 199
Post processing, 211
Principal component analysis (PCA), 76

Q

QA, *see* Quality Assessment
Quality Assessment (QA), 11, 184

R

Radial basis function (RBF), 143
Radiative transfer equation (RTE), 90
RBF, *see* Radial basis function
Red-green-blue (RGB), 71
Reference data, 140–141, 158
Region of interest (ROI), 92
Remote sensing
   -based phenology applications, 81
   with high spatial resolution, 43
   of radiometric surface
      temperature, 91
RGB, *see* Red-green-blue
RMSD, *see* Root mean square deviation
RMSE, *see* Root mean square error
Robust reference dataset creation, 157,
      168–169
   bioregions and VFMP Network in
      Victoria map, 159
   case study, 162–168
   change metrics, 157–158

disturbance attribution, 160
disturbance maps, 165–166
land tenure within case study
    area, 163
mapping disturbance, 165–168
methods, 159
quality control and quality
    assurance, 161–162, 164–165
reference datasets, 158, 161
spectral change signal of forest
    disturbance, 160
study area, 158–159
ROI, *see* Region of interest
Root mean square deviation (RMSD), 31
Root mean square error (RMSE), 34, 60, 91
RTE, *see* Radiative transfer equation

S

Satellite
    LST retrieval algorithms, 90
    remote sensing, 138
    temperature sensor, 90
Scan Line Corrector (SLC), 5, 26, 125, 208
SCD, *see* Sub-annual change detection
Seasonal phenological trajectory, 71
Seasonal time series data, 25
SEVIRI, *see* Spinning Enhanced Visible
    and Infrared Imager
Shape-similarity-match approach,
    12; *see also* Cloud and cloud
    shadow detection
Short-Wave Infrared (SWIR), 11
Single-season phenological analyses, 70;
    see also Phenological inference
    generic trajectory for growing
        season, 72
    per pixel seasonal profiles of land
        covers, 73
    phenological variations, 72–73
    seasonal trajectory, 71
    spectral indicators of phenology, 70–71
    spectral vegetation indices, 71
Single-year phenology, 74; *see also*
    Phenological inference
    agricultural mapping and
        monitoring, 74
    forest mapping and ecosystem
        analyses, 74–76

hydro-phenological analyses of
    complex flooded landscapes,
    76–77
seasonal NDVI contrast in mixed
    deciduous and conifer
    forest, 75
temporal unmixing approaches, 74
SLC, *see* Scan Line Corrector
S-NPP, *see* Suomi National Polar-
    orbiting Partnership
SPARCS, *see* Spatial Procedures for
    Automated Removal of Cloud
    and Shadow
Sparse-representation-based
    SpatioTemporal reflectance
    Fusion Model (SPSTFM), 45
Spatial and Temporal Adaptive
    Reflectance Fusion Model
    (STARFM), 45
Spatial interpolators, 26
Spatial Procedures for Automated
    Removal of Cloud and Shadow
    (SPARCS), 7
Spatial Temporal Adaptive Algorithm
    for mapping Reflectance
    CHange (STAARCH), 45
Spatiotemporal data fusion, 43, 62–64;
    see also Flexible spatiotemporal
    data fusion method; NDVI
    linear mixing growth model
    dictionary-pair learning based
        methods, 45
    methods, 44–46
    for synthetic fine-resolution time
        series, 44
    unmixing based methods, 45, 46
    weighted function based methods,
        44–45, 46
Spatiotemporal interpolators, 27
Spectral indicators of phenology, 70–71
Spectral Vegetation Indices (SVI), 70, 71
Spinning Enhanced Visible and
    Infrared Imager (SEVIRI), 113
Split-window algorithms (SW
    algorithms), 90
SPSTFM, *see* Sparse-representation-
    based SpatioTemporal
    reflectance Fusion Model
SR, *see* Surface Reflectance

SSIM, *see* Structure similarity
STAARCH, *see* Spatial Temporal Adaptive Algorithm for mapping Reflectance CHange
STARFM, *see* Spatial and Temporal Adaptive Reflectance Fusion Model
Structure similarity (SSIM), 60
Sub-annual change detection (SCD), 141, 142
Sub-pixel analysis methods, 122
SUHI, *see* Surface urban heat island
Suomi National Polar-orbiting Partnership (S-NPP), 64
Support Vector Machine (SVM), 14, 122, 126, 139
    modified SVM classification, 143
Surface Reflectance (SR), 177, 178
Surface urban heat island (SUHI), 89; *see also* Annual cycle parameters
    Aqua daytime MODIS LST pattern and SUHI of Paris, 108
    climatological SUHI analysis, 107–112
    for different overpass times, 110
    mean annual SUHI, 109
SVI, *see* Spectral Vegetation Indices
SW algorithms, *see* Split-window algorithms
SWIR, *see* Short-Wave Infrared

T

Tasseled Cap (TC), 18, 125
Tasseled Cap Wetness (TCW), 161, 165
TC, *see* Tasseled Cap
TCW, *see* Tasseled Cap Wetness
Temperature emissivity separation (TES), 91
Temperature sensor, satellite, 90
Temporal interpolators, 26–27
Temporal pattern of global radiation, 105
Temporal unmixing approaches, 74
TES, *see* Temperature emissivity separation
Thematic Mapper (TM), 5, 124, 140, 165, 176, 216
Thermal Infrared (TIR), 5, 90
Thermal Infrared Sensor (TIRS), 5
Thermodynamic temperature, 91

Thiessen Scene Area (TSA), 195
Thin plate spline (TPS), 56
Time integrated disturbance map mosaic, 194
    accuracy for disturbance class, 196
    error matrix, 197, 198
Time series
    adaptive and full length, 144
    of remote sensing data, 70
TIR, *see* Thermal Infrared
TIRS, *see* Thermal Infrared Sensor
TOA, *see* Top-Of-Atmosphere
Top-Of-Atmosphere (TOA), 177, 208
TPS, *see* Thin plate spline
Trajectory-based landscape change analyses, 79–81
Trajectory mapping
    accuracy assessment, 148
    workflow, 141
Trajectory reconstruction, 144
TSA, *see* Thiessen Scene Area

U

UBDF, *see* Unmixing-Based Data Fusion
UDTCDA, *see* Universal Dynamic Threshold Cloud Detection Algorithm
United States Geological Survey (USGS), 4, 26, 92, 176
Universal Dynamic Threshold Cloud Detection Algorithm (UDTCDA), 12
Universal Transverse Mercator (UTM), 191
Unmixing-Based Data Fusion (UBDF), 58
Unmixing based methods, 45, 46
Urban expansion, 205–206
Urbanization, monitoring, 205, 212, 214–216, 217; *see also* Normalized Difference Vegetation Index
    accuracy assessment and evaluation, 211–212
    agriculture land loss in Shanghai, 213, 214
    change detection accuracy, 212–213, 215
    in China, 206

cloud/shadow free NDVI
        composites, 208–209
cookie cutter method, 214, 216
cropland loss in Shanghai, 216
data and preprocessing, 207–208
finding year of change, 210–211
Landsat imagery, 207
methodology, 207
official statistics data comparison, 213
post processing, 211
USGS, *see* United States Geological
        Survey
UTM, *see* Universal Transverse Mercator

V

VCT, *see* Vegetation Change Tracker
Vegetation Change Tracker (VCT),
        81, 175, 176; *see also* Forest-
        disturbance mapping project
    class commission errors, 184
    disturbance analysis, 179
    disturbance map class definitions, 180
    need for Landsat time series, 180–182
    samples of disturbances, 181
    western US sparse forests
        adjustment, 182–184
Vegetation-impervious surface-soil
        (VIS), 125
VFMP, *see* Victorian Forest Monitoring
        Program
Victorian Forest Monitoring Program
        (VFMP), 159
VIIRS, *see* Visible and Infrared Imaging
        Radiometer Suite
VIS, *see* Vegetation-impervious
        surface-soil

Visible and Infrared Imaging
        Radiometer Suite (VIIRS), 64
Visual assessment of samples of
        disturbances, 181

W

WA, *see* Wavelet Analysis
Water vapor absorption band, 8
Wavelet Analysis (WA), 18
Web-enabled Landsat Data (WELD), 191
Weighted function based methods,
        44–45, 46
Weighted Linear Mixing (WLM), 51
WELD, *see* Web-enabled Landsat Data
WGS-84 UTM, *see* World Geodetic
        System Universal Transverse
        Mercator
WLM, *see* Weighted Linear Mixing
World Geodetic System Universal
        Transverse Mercator (WGS-84
        UTM), 124
World Urban Database And Portal Tools
        (WUDAPT), 111
Worldwide Reference System (WRS),
        123, 162, 175
WRS, *see* Worldwide Reference System
WUDAPT, *see* World Urban Database
        And Portal Tools
Wuhan city, 123, 124

Y

YAST, *see* Yearly Amplitude of Surface
        Temperature
Yearly Amplitude of Surface
        Temperature (YAST), 91

T - #0223 - 111024 - C0 - 234/156/13 - PB - 9780367571795 - Gloss Lamination